Bioinformatics and Systems Biology

Collaborative Research and Resources

Frederick B. Marcus

Bioinformatics and Systems Biology

Collaborative Research and Resources

Springer

Dr. Frederick B. Marcus
Principal Scientific Officer
Research Directorate General
European Commission
1049 Brussels
Belgium
Frederick.Marcus@ec.europa.eu

DISCLAIMER: The contents of this book are based upon referenced, publicly available sources, specifically books, publications and websites. Although at the time of publication, the author is an employee of the European Commission, this book is his work alone and it is not sponsored by the Commission, nor is it a Commission publication. The author is not receiving any royalties on this book. The contents may not in any circumstances be regarded as stating an official position of the Commission. Neither the Commission nor the author nor any person acting on behalf of the Commission is responsible for the use that might be made of the contents of this book. Material in this book is only an indicative guide to accessing the officially approved material available in Commission websites and publications.

ISBN 978-3-540-78352-7 ISBN 978-3-540-78353-4 (eBook)
DOI 10.1007/978-3-540-78353-4

Library of Congress Control Number: 2008921486

Cover design: WMX Design GmbH, Heidelberg, Germany

Printed on acid-free paper

9 8 7 6 5 4 3 2 1 0

springer.com

I dedicate this book to my parents Marvin and Aileen Marcus, my Wife Rosemary and my Brother Jeff and my departed grandparents Jennie Marcus and Reuben and Anne Axler, who gave me the greatest gift, love, and all that goes with it.

Preface

Purpose of This Book: This textbook on collaborative research in bioinformatics and systems biology, which are key elements of modern biology and health research, highlights and provides access to many of the methods, environments, results and resources involved, including integral laboratory data generation and experimentation and clinical activities. Collaborative projects embody a research paradigm that connects many of the top scientists, institutions, their resources and research across Europe and the world, resulting in world-class contributions to bioinformatics and systems biology.

Central Themes: A number of themes are expressed and described, which guided the selection of material and its presentation:

- This book concentrates on collaborative research projects which have a significant computational biology component. A chapter with a title such as "Cancer" therefore covers the area in this restricted context. Moreover, most large-scale collaborative projects in Europe are funded by the European Commission. These projects often unify the best laboratories in Europe, which in turn are often themselves linked to worldwide programmes. Therefore, the research tends to accurately reflect the most up-to-date and best state-of-the-art research, which usually has a computational component.
- Computational approaches are a key part of much of modern life sciences research. This book aims at making researchers aware of the central importance of bioinformatics and systems biology in modern life sciences research and the wide range of publicly available resources generated by collaborative research programmes.
- A guide is needed for researchers to access the full range of resources available. Researchers are aware of laboratories, publications, databases and tools related to their areas, such as the European Bioinformatics Institute (EBI 2007) of the European Molecular Biology Laboratory (EMBL 2007), but in the form of individual pieces of a puzzle that they need for their work. They are not aware of how these resources are being linked together, nor what has been accomplished

by doing so. This book shows how collaborative researchers are putting many of the pieces together in ways accessible to the entire biomedical community.

- Collaborative research approaches are highly productive and often essential. Extensive multilaboratory collaboration is necessary for assembling the scale of resources needed to advance in many areas depending on computational biology, especially when closely linked to experimentation. European collaborative research is highly successful owing to the autonomy and flexibility given to the researchers. Tools and resources have been assembled and developed which cover much of modern biology, ranging from gene definition and alternative splicing to protein sequence, structure, function and interaction networks, with direct application made to disease processes. Many similar projects are interlinked, leading both to broad resource development and to interrelated research programmes.
- Collaborative research results involving computational approaches represent the state of the art in many areas. This book aims to describe the most advanced research results and resources available, and to make them as accessible as possible in the form of a textbook and user manual. Often the best individual resources and results are mobilised for the best collective research.
- The research and resources described in this book are of worldwide interest and relevance. Even though the projects described are mostly funded by the European Commission with predominantly European participation, these projects are strongly interactive with worldwide resources. The resources involved include access to the databases EMBL-Bank for genome sequence, UniProt for protein sequence, Ensembl for genome browsing, MSD for protein structure, etc. Therefore, even though many tools originate from European projects, gateways are provided to worldwide research and resources.
- Science management is a key element of collaborative research. Many textbooks teach the underlying science, tools and procedures necessary to carry out research, but very few discuss how to plan and carry out a research programme, especially at the collaborative level. Each stage of a project, from planning to proposal to project organisation to project operation, requires optimal organisation and structures for optimal success. This book serves as a guide to understanding methods of modern collaborative research, and to assembling the level of resources needed for the complexity of much of modern life sciences research.
- The European Commission plays a key role in creating a collaborative research environment. Another motivation is to illustrate the role of the European Commission in health research. The Commission's health research budget is smaller than the total of that of the member states of the European Union, but it is used strategically to beneficially link resources together, and is doubling under the new Seventh Framework Programme for Research (FP7 2007) compared with the Sixth Framework Programme (FP6 2007).

Structure of This Book: Following the Preface, Contents and introductory chapter, the book is organised in four parts which are somewhat analogous to the so-called central dogma of molecular biology: *sequence*, *structure*, *function*, *phenotype*. Part I (fundamental collaborative research and computational biology) shows the *sequence* of research approaches that integrates various elements of the "central dogma" and much more besides, via bioinformatics, systems biology and developmental biology approaches. Part II (resources supporting bioinformatics and systems biology research) discusses the data and computational *structures* for research that have been created, and those infra*structures* needed to generate the data. Part III (disease-related collaborative research and computational biology) exploits the *function* of the research and tools to study infectious and major diseases, including cancer. The chapter on genetic variation and diseases explores one of the great challenges within the "central dogma", how to integrate all the resources on germ-line and somatic genetic variation into disease research. Finally, Part IV (science management, perspectives and conclusions) explores the overall *phenotype* of research itself, what it looks like and how it is organised, its perspectives and outstanding results.

Information Available: This book provides a snapshot of much of the current state of the art in bioinformatics and systems biology research. It is also a practical guide aimed at students, academic and industrial researchers and managers in life sciences and medical research, with information and pointers to resources. Most of the results and resources described are available worldwide through the Internet and international grid connections, and link to most of the major worldwide databases and tools.

Others besides researchers will find extensive sections of this book useful. Much of the introductory chapter and the chapter on science management is intended for the general reader, and give insights into collaborative research in general, and how it is supported in particular by the European Commission. A valuable feature of this book is that it shows how research is planned, organised and carried out in a variety of areas, in contrast to books that concentrate only on the science.
Specifically, the book discusses:

- Collaborative research paradigms
- Scientific basis and current state of the art in bioinformatics and systems biology, and their applications to disease processes
- Key scientific results and ongoing research
- Resources and infrastructures created by the projects
- Practical guidance to project and related websites and software and services
- Sources in books and the scientific literature
- Methods for accessing the knowledge and linking to existing projects
- Practical information about creating and participating in collaborative research projects
- Future perspectives

How To Use This Book: This book is intended to act as a guide for life sciences and biomedical researchers to the research and resources being developed by European Commission collaborative programmes, and to the individual laboratory resources that they link together. There are several ways of finding information:

- Table of contents: The table of contents is presented as a three-level detailed table of contents for finding individual research and resource areas.

- Summary tables and lists: Tables and lists are presented in the Chap. 1 that provide access to project websites and their participants and publications. Lists are also provided of project catalogues for the whole range of health research, and to relevant resources.
- Index: The Index provides single-phrase and word access to discussions of key scientific areas.
- References and access to websites: The main access points are over 350 websites listed in References along with over 170 key published papers by the projects discussed and supporting reference books. Reference names often correspond to project titles, and give direct access to their website home pages. These websites are often gateways and portals to many relevant tools and capabilities and databases. The websites themselves are vast reservoirs of information, with various forms of documentation, and extensive lists of journal publications resulting from project activities. The website documentation provides more information about the research process itself and the history and means of developing tools than may be found in the literature or instruction manuals. This book attempts to make that knowledge accessible, showing the resources available and their organisation.
- Reference forms: A text reference such as BioSapiens (2007) refers both to the project called BioSapiens and to the BioSapiens website, with the address listed in the References. The "2007" following the reference name indicates that it is a reference rather than just a project name. The date "2007" for such references means that the website was recently accessed in 2007 and is currently available, even though it may contain material from a variety of dates which may be earlier. All websites were verified on May 2008.
- Access for non-specialists: Non-specialist but scientifically oriented readers may wish to concentrate on chapters in the following order, skipping some technical sections: Chap. 1; Chap. 12, Chap. 11, Chap. 10, followed by the "Introduction" sections of the remaining chapters.
- Navigation through European Commission websites: Chap. 10 refers to various websites of those involved in the various stages of European Commission collaborative research programmes, including proposal preparation.

Brussels, Belgium *Frederick B. Marcus*
May 2008

Acknowledgements

I greatly appreciate the comments and contributions and reports from the leaders of the projects I have supervised, especially from Jozef Anne, Rolf Apweiler, Terri Attwood, Ewan Birney, Alvis Brazma, Sierd Bron, Anthony Brookes, Soren Brunak, Graham Cameron, Fabio Fiorani, Daniel Gautheret, Les Grivell, Colin Harwood, Kim Henrick, Henning Hermjakob, Ralf Herwig, Pierre Hilson, Cees van den Hondel, Pascal Kahlem, Martin Kuiper, Hans Lehrach, Jack Leunissen, Alberto Luini, Philippe Noirot, Kerstin Nyberg, Josep Roca, Karsten Schurrle, Luis Serrano, Janet Thornton, Anna Tramontano, Alfonso Valencia and also external reviewers Stefan Hohmann, Olaf Wolkenhauer and Boris Zhivitovsky, and the editors Sabine Schwarz (formerly Schreck) and Ursula Gramm of Springer-Verlag, Copyeditor Stuart Evans, and T. Saravanan of SPi. I also gratefully acknowledge the support of my many colleagues at the European Commission.

Contents

Chapter 1
Introduction

Abstract This book provides a state-of-the-art description of research in bioinformatics and systems biology as supported by European Commission collaborative research projects and functions as a practical guide to project websites, participants, coordinators, publications and key sources. In addition to an introduction to the science, a number of links to key websites and information resources are presented. This introductory chapter illustrates that bioinformatics and systems biology consist of more than databases and tools, and are scientific areas in themselves and major components of most research projects. The key domains for analysis and research are described in detail. Aspects of collaborative research projects are explored, and it is shown how they constitute an important research paradigm.

Introduction

Bioinformatics research involves more than the alignment of DNA sequences, and systems biology involves more than a pathway diagram of the Krebs cycle/citric acid cycle (Krebs 1953). Bioinformatics databases and analysis tools are indispensable parts of most life sciences research projects. Increasing numbers of projects take a "systematic" or a systems biology approach. This chapter outlines the scope and breadth of these related and overlapping fields, which include computational biology and much more besides. It describes how the European Commission has strengthened these areas and expanded them in new directions empowered by the paradigm of large-scale collaborative research. Finally, it indicates how this book can be used as a textbook for finding and using the tools and resources developed, and as a route map to the state of the art of modern biology and health research.

F.B. Marcus, *Bioinformatics and Systems Biology*,
DOI 10.1007/978-3-540-78353-4,

Bioinformatics

What Is Bioinformatics?

In a series of European Commission workshop reports, see Table 1.1, the status of and trends in modern bioinformatics were analysed, and are fully updated here. Bioinformatics derives knowledge by computer analysis of biological and molecular biology data. It is a rapidly growing branch of biology, highly interdisciplinary, and it uses techniques and concepts from informatics, statistics, mathematics, chemistry, biochemistry, genetics, physics, linguistics and other fields. The biological data can be the information stored in the DNA sequences, experimental results from various sources, three-dimensional protein structures, gene expression arrays, patient statistics, scientific literature, etc. An important part of research in bioinformatics is the development of methods for storage, retrieval and analysis of these data. The concept of "information in the genetic code" has its limitations. DNA can be analysed both as a text and as a molecule that interacts with a variety of other molecules. Interactions with and among proteins are governed by three-dimensional structures and their dynamics and flexibility. These in turn are obviously determined by the sequence of bases, but the behaviour of proteins cannot be fully described by reducing its analysis to a one-dimensional level. The three-dimensional aspect is also crucial in understanding protein sequences and protein structures.

References

Introductions to the basis and use of bioinformatics tools and concepts may be found in many textbooks (Attwood and Parry-Smith 1999; Baxevanis and Ouellette 2001; Claverie and Notredame 2003; Higgins and Taylor 2000; Lesk 2001, 2002; Mount 2001; Orengo et al. 2002; Pevsner 2003; Sensen 2006; Polanski and Kimmel 2007; Lengauer 2007; Nagl 2006). Surveys of the current state of various aspects of bioinformatics and systems biology and future prospects are available in the reports on the European Commission workshops shown in Table 1.1.

Table 1.1 Bioinformatics/systems biology relevant European Commission workshops

Workshop description	Reference
Intellectual property rights and bioinformatics and the influence of public policy	Crespi (2001)
Bioinformatics – structures for the future	Gyorffi and Marcus (2003)
Computational systems biology (CSB) – its future in Europe	Marcus et al. (2004)
EU projects report on systems biology	Jehensen and Marcus (2005)
Future needs for research infrastructures in biomedical sciences in Europe	Faure et al. (2005)
European database and analysis resources for research in human genetic variation	Marcus and Mulligan (2006)
Infrastructure needs for systems biology	Cassman and Brunak (2007)

Trends in Bioinformatics

Bioinformatics gained significant programmatic importance during the European Commission's Fifth Framework Programme (FP5 2007) from 1998 to 2002, and in the wider scientific community, where activity in this field was mainly related to development of storage and organisation of the growing amounts of data produced by ever more sophisticated genetic technologies, in conjunction with the infrastructural needs accompanying basic genetic research. In the Sixth Framework Programme (FP6 2007) from 2002 to 2006, bioinformatics expanded to the development and use of computational tools for the biological interpretation of the large amounts of data. There is a very heterogeneous scientific community that covers all aspects of today's genetic research. Europe is engaged in several international collaborations, the main partners being the USA and Japan, and others as well. In the Seventh Framework Programme (FP7 2007) from 2007 to 2013, the commitment to bioinformatics and systems biology will be extended even further with increased funding and explicit commitment to both of these areas.

What Is the State of the Art?

Key areas under rapid development include:

- Comparative DNA and protein sequence analysis tools, gene finding and genome browsers
- Identification of functional non-protein-coding sequences, e.g. regulatory sequences
- Imaging databases
- Integrated sequence-based functional data from microarrays, RNA interference
- Human genetic variation data (e.g. single-nucleotide polymorphisms, haplotype map, sequencing)
- New literature and text mining tools
- New "-omics": transcriptomics, metabolomics, lipidomics, glycomics, interactomics, spliceomics, reactomics, etc.

The Role of Bioinformatics in Studying Health and Disease

There is extensive research performed in this field. A major problem in this research, but present more widely, is the tendency to store a minimal amount of data representing successful outcomes only, usually in processed form. More should be permanently conserved. The more medically and patient-orientated the research, the more confidentiality becomes a key issue. When bioinformatics reaches the level of dealing with personal data there are legal and ethical considerations that become apparent. Even in the intellectual property rights protection oriented world of pharmaceuticals, open-source licensing is a valuable approach to encourage maximum transfer and use of data. Immunological bioinformatics is needed in the vaccine design area and in problem-driven cases, e.g. asthma. Access to and understanding of distributed, heterogeneous information resources is critical but it is a complex, time-consuming process, because of thousands of relevant information sources. Rapidly changing domain concepts and terminology and analysis approaches confound the situation, as do constantly evolving data structures, continuous creation of new data sources and highly heterogeneous sources and applications. Data and results are of uneven quality, depth and scope. Collaboration for understanding and consensus is essential. One of the major contributions of European collaborative research has been to address and partially solve these problems.

Bioinformatics Within Life Sciences Research Projects

Bioinformatics now plays a key role in the integrated design of experiments. Genomics-based life sciences projects now recognise that significant resources should be devoted to bioinformatics, for planning experimental procedures and the analysis, interpretation and publication of results, as well as for storage of data and results in accessible and comprehensible form. Laboratories are already hiring their own local bioinformatics experts to keep pace with the growing set of available techniques. This detailed work is only possible thanks to the availability of effective methods on the World Wide Web, reliable service providers and the proximity of other bioinformatics experts. The way in which biologists address specific problems is often influenced by the availability of bioinformatics methods for the design, management and interpretation of the results. This book aims to greatly facilitate that process.

Types of Analysis

In the fields of genomics and proteomics, bioinformatics provides the key connection between all different forms of data gathered by new high-throughput techniques such as systematic sequencing, proteomics, expression arrays and yeast two-hybrid. We have at our disposal hundreds of genomes, thousands of protein structures, protein interactions determined by yeast two-hybrid and tens of thousands of genes with their expression monitored in hundreds of experiments, and millions of single-nucleotide polymorphisms. Handling this massive amount of data requires powerful integrated bioinformatics systems. Issues related to database interoperability, information representation and data description (the much-abused term "ontology") are currently being addressed. In fields such as automatic extraction of information from the biological literature, activity has increased greatly. Further development of data analysis algorithms is necessary to address the many unsolved problems. Algorithm development is becoming biologically oriented, although physics and engineering approaches are also essential. Networking is necessary to foster integrative approaches and prevent the development of many solutions to the same problem.

Ontologies

An ontology is a system of coding knowledge in such a way that it is computer-readable. Ontologies are central to classification and functional descriptions, and biological interpretation always needs a biological context to remain meaningful. Function does not have a universal metric and cannot be described except in the process in which a molecule is involved. According to the context, a single molecule can have many or even innumerable functions (e.g. water). Gene Ontology (GO 2007), is a much-cited piece of software, developed by Michael Ashburner and others, which is a dynamic controlled vocabulary that can be applied to all organisms, even as knowledge of gene and protein roles in cells is accumulating and changing. The three organising principles of Gene Ontology are molecular function, biological process and cellular component.

Genomics and Proteomics

The ability to predict gene products from genome sequence is rapidly changing. A key goal is to identify gene products (RNA and protein) that can actually be detected in cells. It is necessary to find all the encoded molecules in vivo and in vitro, as there are limited prediction capabilities *in silico*. Many of the interpretative problems are still unsolved. Genomics was initially driven by data collection and much interpretation is hypothesis-driven. A new direction is to provide simple tools so as to be able to access databases and perform analyses to build generalised models for biological processes. The best approach is comparative genomics leading on to functional studies with improvement of sequence analysis tools and text mining tools and new solutions for knowledge representation. Gene regulation is a key area and needs well-defined experiments and a computational framework. Sequence data are extremely valuable and new sequences are very good value for money, especially with sequencing costs falling rapidly. The added value of new genomes is important. A sequence is a needed basis for many modern experiments and the extensive use of UniProt (2007) shows what a valuable resource this is to catalogue the proteins. Newly sequenced genomes do give additional information beyond that of the organism itself by increasing the value of previous data, e.g. regulatory elements, and comparison can lead to fuller understanding. Haplotype analysis for disease identification is a key element for developing understanding of the disease processes, with personalised treatment a key goal. Haplotype information can be gained by repeated genomic sequencing and is incorporating sophisticated analysis tools to find associations with disease.

New instrumentation, high throughput, automated, particularly nucleotide sequencing and arrays, is enabling biologists to generate vast quantities of information. The global databases are collaborating with one another, sharing the workload of receipt, curation, annotation and storage. There remain unresolved subtleties in detecting genes and regulatory elements within the total genome. The rapid development and diffusion of arrays is producing new challenges. The biologist is in danger of drowning in this flood of data, but projects are in progress to deal with this. Experiments are not readily replicable, especially not when conducted in different places, by different experimenters, with different arrays and without standard operating procedures, etc. Science becomes gravely hampered if scientists in different locations or contexts and at different times are unable to communicate data and results reliably to one another – similar problems are facing curators of databases, public or private, hence the result of developing common languages and protocols in ArrayExpress (2007).

Structural Genomics

Homology modelling is the most successful tool for understanding the significance of protein three-dimensional structures. Part of the structural genomics effort goes into covering the protein structure space sufficiently well that "all" proteins can be modelled with sufficient accuracy. Better homology modelling will reduce the number of structures that need to be solved experimentally. The quality of homology models is primarily determined by the accuracy of the alignment between the sequences of the query protein and the template structure and the force field that is used in the modelling. For low to very low sequence similarity, homology models can only give qualitative features of the protein. Function prediction from the structure alone remains an elusive goal and one way is to search for conservation of structural templates. Aspects of function are investigated by looking at physical properties that can be calculated from the structure, such as shape, electrostatic potential and molecular dynamics. An important part of structural bioinformatics, in particular at a medically oriented research institute, is the study of the interaction between a protein and ligands, building up drug design and virtual screening activities. The holy grail of molecular biology is the idea that sequence can predict structure and that structure can predict function. However, this is true only to a limited extent, given the multiple protein forms from alternative splicing and post-translational modification, and the multifunctional nature of many proteins once they have been formed, e.g. degree of phosphorylation. It is therefore necessary to identify all the parts (molecules) to understand the whole (cell and organism). The reality is that sequence, structure and function need to be determined and related in order to gain understanding and build all the parts into models of working systems.

Systems Biology

What Is Systems Biology?

Systems biology involves developing the understanding of a biological system through mathematical and computational modelling of the interactions of components of the system, leading to the expression of this understanding in qualitative and quantitative terms – in particular, in terms amenable to electronic storage and communication. Examples of "biological systems" are as old as modern biology, e.g. the Krebs (1953) cycle in metabolism. However, the Krebs cycle represents an example of what arduous enzyme-by-enzyme approaches can accumulate over time. These no doubt are systems, but they are not accomplished by a systems biology approach, nor do they constitute modern systems biology. Once this knowledge is framed in a dynamical simulation model it can, however, be used for systems biology. Systems biology approaches have long been used in physiology and in modelling the effects of medicines, as pharmacokinetics and pharmacodynamics. The key changes that make a modern approach to systems biology necessary and possible are the very recent developments of high-throughput technologies in many domains of biology as well as emerging technologies that allow the generation of new types of quantitative data at high precision and resolution. Furthermore, recent developments of complex systems theory have provided us with the mathematical concepts and tools needed to understand some of the dynamical phenomena observed in the living world. Analysis of just a small fraction of the available data has led to the realisation that understanding of biology, health, disease and medicines requires an integrated approach to studying the processes involved. Even with our current state of knowledge, systems biology has already made impressive contributions to both fundamental understanding and to direct applications to health. As an example, circadian rhythms can only be fully described with a systems biology approach. Carefully tailored models can already make useful predictions about disease processes and how medicines can be optimally applied, constituting the beginnings of personalised medicine. For example, the time of day may be optimised for medication with chemotherapy or for diabetes treatment.

References

The field of systems biology is evolving rapidly. Excellent textbooks are available on the history and methods of systems biology, on its computational and experimental aspects and on available models and databases (Kitano 2001; Fall et al. 2002; Alberghina and Westerhoff 2005; Alon 2006; Klipp et al. 2005; Stelling 2004) and on applications (Bringmann et al. 2007; Bertau et al. 2007). An extensive summary of the worldwide state of the art and institutions, e.g. ISB, has been compiled by Cassman et al. (2007). Within the European Commission, systems biology workshops reports (see Table 1.1) provide a useful perspective on the state of the art and on future developments.

Trends in Systems Biology

Computational and experimental biology have for many decades been separate disciplines. Systems biology, on the other hand, emerged as a new discipline in which theoreticians and experimentalists closely collaborate, ideally from the planning of an experimental study. There is a need for a continuous and iterative collaboration between modeller and experimentalist such that the modeller understands biological knowledge about the system and takes part in the definition of new experiments and the experimentalist understands the principles of converting biological information into mathematical descriptions. The need for this close interaction partly arises from the lack of databases with sufficient information for modelling, and is in contrast to many types of bioinformatics analysis which can be based on well-structured and comparatively simpler types of data, such as DNA sequences.

A dominant theme is constructing complex systems from genes to cells by combining knowledge from different databases, different types of data, etc. This approach has been used with some success, but generally it creates more questions than answers. Scenarios emerging from such approaches are sensitive to the precise way the systems are built. Data are important, but the way these data are utilised in modelling is much more important. In academia it is of interest to develop complex models based on genetic information, etc., and many "virtual cells" show interesting cell-like behaviour. However, industry has special problems with validation. An industry staking hundreds of millions of euros on a medicine or bioprocess based on a computational model needs that model to be as useful, accurate and comprehensive as possible. This is not achieved by better data alone. Rather, the way the model is built, the intimate interplay between its parts and even the software used to make the calculations are critical. This also means that tool development in systems biology is critical and needs to be both practical and visionary. Systems biology profits from many different disciplines in the life sciences as well as bioinformatics, information technology, dynamic systems theory, etc. Systems biology is a science in its own right in that it aims at discovering unknown principles and "laws" that occur in biological systems.

What Is the State of the Art?

Impressive advances have been made in modelling a number of individual processes in physiology, e.g. in modelling the dynamics of the heart (Noble 2007). Increasingly, however, modelling of molecular processes, involving most or all genes, gene products and metabolites is being used to understand complex disease processes. However, it is vital to appreciate where models have worked, and where not.

The most successful current implementations of systems biology rely on iterative cycles of data analysis and computerised (*in silico*) model construction/refinement and predictions, linked to wet-laboratory (in vitro) and living specimen (in vivo) experimental design, experimentation, and data capture and storage in forms that can be represented and manipulated by computer software.

Why Do Systems Biology?

A systems approach to life sciences research is often necessary for better qualitative and quantitative understanding of the functioning of biological systems in healthy and pathological conditions, especially where multiple time-dependent pathways are involved. This includes the exploitation of biological information generated by high-throughput technologies, as well as new types of data becoming available through developments in a broad range of experimental techniques such as mass spectroscopy, enhanced Raman spectroscopy, photoluminescence, space- and time-resolved laser interference, automated patch clamping, nuclear magnetic resonance spectroscopy, bioimaging, etc. The current dominant research paradigm of one laboratory–one gene–one function–one disease has reached its limits in many areas, as diseases or important traits often arise from a combination of factors. Cataloguing and classification of molecules will ultimately not suffice to reason about the function of molecules or functioning of cells. Many important aspects of biology can only be researched by a combined experimental and computational approach to developing systems models that yield useful results, for example when examining interlinking pathways. The implications of the findings in the life sciences will be immense and will allow a virtual description of cellular and physiological activities and functions. The ultimate goal is to understand the functioning of physiological and pathophysiological systems, and thus to rationalise model-driven drug development, and to investigate effects of treatments or support fruitful approaches in biotechnology. Knowledge gained from systems biology may also open up disease prevention and treatments avoiding extensive use of drugs. Oxford University professor Denis Noble (2002) says: "Successful physiological analysis requires an understanding of the functional interactions between the key components of cells, organs, and systems, as well as how these interactions change in disease states. This information resides neither in the genome nor even in the individual proteins that genes code for. It lies at the level of protein interactions within the context of sub-cellular, cellular, tissue, organ, and system structures."

Approaches to Modelling

There are two main modelling approaches in systems biology (also computational/ computational systems biology). Besides the construction of large-scale models, incorporating as many details as have been uncovered experimentally on a given pathway or signalling system and on a detailed "cartography" of various networks, another useful approach in systems biology relies on the construction of small-scale models of limited complexity, containing a reduced number of variables (two to 20), and aiming at addressing specific questions. From these small-scale models, one can often derive conclusions of more general significance, e.g. concerning cellular rhythms, cell signalling and cell cycle dynamics, especially when including dynamic phenomena: multistability, oscillations, spatial and spatial–temporal patterns (e.g. in morphogenesis and cell-to-cell communication). Both approaches have merits and limitations, and they can converge by putting small-scale models (modules) into a common framework. Model standardisation is vital, but should not be used to suppress creative approaches. Models should also be closely tied with existing and/or new experimental data. Within the realm of smaller-scale models, there is the further division into a numerical simulation approach, by far the most common, and a mathematical solution approach, which is especially capable of giving important insights into complicated problems when the key features can be identified and quantified. A full analysis and comparison of the various models, their standards and their tools is provided by Klipp et al. (2007). Websites are available which access a wide range of systems biology models (BioModels 2007; Sysbiomodels 2007).

Who Will Benefit from This Approach?

There are many "customers" that will benefit from systems biology, all of whom/which will ask different types of (scientific) questions. They include researchers in biology, physiology and medicine, in universities, research establishments and the pharmaceutical industry, hospital clinicians, family and research doctors, patients, public health practitioners, the pharmaceutical industry, new and existing biotechnology enterprises, agro biotechnology, etc., and scientists in particular disciplines, e.g. cellular biology, physiology and medicine. Expected benefits will be widespread, since a thorough understanding of complex biological processes will allow tackling many real-world problems. In providing solutions to many of these problems, systems biology might therefore be one of the key approaches of the twenty-first century.

Systems Biology Within Life Sciences Research Projects

Proposals that claim to have a systems approach should contain a strong analysis and modelling component. Systems biology related projects should assign an adequate fraction of resources for data modelling or integration, depending on the scope of the project. Even smaller projects need significant data integration and computer modelling resources, going well beyond basic data analysis. This may encompass projects not only with medical/health goals but also projects in biotechnology such as genetic or physiological engineering of microorganisms or plants or process improvements.

Types of Research Outputs

There are several dimensions in the types of output desired, for example the different levels in the paradigm of doing science: explain, discover, predict, control:

1. Explain.
   - Simulate systems to show that data are internally self-consistent and sufficient for testing hypotheses.
   - Perform functional genomics analysis of expression data to study systems under normal conditions.
   - Test hypothesis of how pathways operate.
   - Use model-assisted data mining to increase understanding.
   - Elucidate properties and (conserved) rules according to which biological systems operate.
2. Discover.
   - Study models of systems *in silico* in order to discover laws and principles for the behaviour of biological systems.
   - Use analytical mathematics to prove those laws and principles.
   - Examine experimental systems so as to assert the relevance of the laws and principles.
3. Predict.
   - Design wet-laboratory experiments to explore certain hypotheses.
   - Perturb computational systems in various ways to see how the system would react to different stimuli, either minor via concentration changes, or major, e.g. by knockout of a whole gene.
   - Simulate and experimentally test the effects of new stimuli (drugs) or other perturbations on a physiological or cellular system, predict the response of patients to different types of treatments.
4. Control.
   - Use models to design new stimuli (drugs) or other perturbations to achieve desired effects in physiological or cellular systems.
   - Control the generation of new data to be able to decide between models, and/or to improve parameter estimates, based on automated analysis of alternative models.
   - Develop a range of industrial applications for controlled biological systems.

Methods

Many of the successful models are oriented to answering particular questions. Up to now, no model incorporates all data, solves everything or is available to answer all questions. Many successful models in fact have the following characteristics:

- They are strongly coupled with experimental work occurring in conjunction with the modelling.
- They tend to ask one question and often focus on the time dependent behaviour of one parameter, even though the modelling may be much more complex.
- Many successful models in the literature solve between two and ten differential equations, and often employ arbitrarily set parameters. This already involves enormous simplifications, since real biological networks are often much more complex. The availability of both much larger datasets generated as parts of functional genomics projects and – perhaps more significantly – far more accurate measurements of specific processes, as well as new modelling techniques, could very well change this situation, and allow modelling to proceed on a more realistic scale.
- There are successful modelling alternatives to the use of differential equations, for example Boolean networks, which are not as detailed, but which can answer many useful questions.
- Multiple-level modelling is being attempted, but the modelling at each level is strongly focused on the particular problem to be solved.
- Successful modelling depends on having a wide range of biological data available in a consistent and quantitative form, although only a small fraction of it may be used in a particular model.
- More and more modelling is done using standard platforms for programming such as Systems Biology Markup Language (SBML 2007), involved in over 100 software systems, powered by standard databases such as Reactome (2007) and KEGG (2007). However, the data and the choice of modelling are often influenced by a coupled experimental programme.
- Modelling also needs to invoke complex systems theory. Biological systems operate far from thermal equilibrium, and display self-sustained oscillations, pulses and bursts, at all different levels of organisation. The application of bifurcation theory has for the first time provided us with the mathematical tools and concepts needed to deal with such problems. Yet, many mathematical problems in this field still remain unsolved. We do not understand, for instance, how a simple model of a bursting and spiking cell can change from one type of dynamics to the other. The interaction of many oscillating subsystems (cells) with slightly different parameters is also a matter of great importance.
- For systems biology collaborative research consortia, it is necessary to work under standardised conditions, especially if more than one team or project is employing the same model organism. Working with the same standard operation procedures will ensure comparable results; experimental conditions as well as data analysis and presentation have to be well documented since incomplete information is useless. Consortia focus on well-defined biological questions and tackle those in one or a limited number of different (model) organisms to ensure coherent results of wide impact. Since systems biology is an interdisciplinary research area, modelling efforts include experimental work (wet laboratory) and method development.
- Models should make some attempt to connect with functional genomics, in the sense of ultimately contributing to understanding the entire living organism.

Research Areas for Analysis

Dimensions of Life Sciences Research

A structure for analysing life science research areas is provided by McCulloch AD, Huber G, Integrative biological modelling in silico: 4–19 in Bock et al. (2002), who describe the dimensions covered by systems biology research and its anticipated development, e.g. the integrative analysis of physiological function. Using the computational models of the heart as examples, they discuss three types of integration: structural integration, functional integration and synthesis. Systems biology covers several other orthogonal dimensions of integration and scale of analysis as well:

- Structure: gene, protein, macro complex, organelle, cell, network, tissue, organ, organism
- Function: regulatory, growth, metabolic, electrical, mechanical, transport, signalling
- Data to theory: empirical data, statistical modelling, predictive modelling, mathematical, ontologies, systems analysis, physical–chemical principles, theory
- Choice of model organism (and advantages for studies) or man: bacteria (rapid reproduction), monocellular eukaryotes (still relatively simple, rapid reproduction), mammalian cells (laboratory cell lines), fertilised eggs (developmental studies, but more difficult to obtain, maintain, artificial environment), model organisms (complex, more relevant), humans (highly complex and relevant, ethical and legal restrictions)
- Systematic to systems biology: time dependence, degree of connectivity of networks, interplay of experiment and computation, system perturbations, predictive capacity
- Reproductive and developmental processes of organisms
- Evolution and diversity at the population level
- Cross-organism interaction (disease interactions and transmission, food chain transmission, development of bacterial resistance, mutants of HIV, new influenza variants, etc.)

Choice of Area

Since bioinformatics and systems biology research has such a wide range of "dimensions", when deciding on research areas, how can researchers choose the most promising area or regions in this multidimensional space? Often the choice is made by building on previous work and capabilities at individual laboratories. The challenge for the future is to identify where bioinformatics and systems biology methods can be applied to usefully improve knowledge and applications, often expanding analysis to higher levels of integration of data. In many cases, the greatest insights are still obtained by "old-fashioned" direct observation of the living organism in the absence of modelling, since modelling sometimes just reproduces a limited data result, but does not bring new insight. Luckily, because of dominant mechanisms and homeostasis, it is possible to create useful reductionist models at several levels of complexity and biology, which are capable of simulating and understanding existing data, with different model complexity:

- Bioinformatics (genotype–phenotype)
- Systematic biology (pathways with linear linkage)
- Systems biology (single or interlinked pathways with quantitative calculations iterated with experiments)
- Physiology with key effects identified (e.g. heart modelling)

"Right-Size" Science

Because of the inherent complexity of living systems, biologists are necessarily reductionists, from the "small-science" doctoral student looking at one DNA sequence to large international "big-science" teams trying to develop computer models of multilevel physiological systems. The challenge is to find the "right-size" science level for each problem. As experience has already shown, nature had provided us with abundant opportunities to do just that. Different approaches have been successful, which ultimately need to be merged: data-driven modelling using very large databases; top-down model-driven, and fully analytic approaches.

Collaborative Research

The Nature of Collaborative Research

By describing the state-of-the-art research being carried out and planned, this book will demonstrate that partly owing to these collaborative research projects, European research and resources are among the world's best in many areas of bioinformatics and systems biology. This book also functions as a route guide and manual for collaborative research, particularly in the European context. As a guide, it is relevant to the entire world, not just Europe, since most results and resources are available worldwide through the Internet and international grid connections, and researchers in effectively all countries can collaborate with or participate in European research programmes. European collaborative research is highly successful and functions extremely well, particularly in these areas of computational biology using Web-based communications, owing to the autonomy and flexibility given to the researchers through the proposal, contract and project management policies of the European Commission. This extensive multilaboratory collaboration is often essential for assembling the scale of resources needed to advance in the fields discussed here. Muldur et al. (2006) provided an excellent description of how the Framework Programmes are actually developed at the political level and implemented in a wide range of research areas. See also ERA (2007) Important collaborative areas have also been established in several other biology and health research fields not extensively discussed here. One example is PRIME (2007). Several other areas are summarised in the project book of the fundamental-genomics group (Fundamental-Genomics 2007) in FP6 (2007).

Advantages of Collaborative Research

The emphasis in this book on an in-depth examination of collaborative research distinguishes it from other bioinformatics and systems biology texts which describe the elements or the working of the pieces of the puzzles that face life sciences researchers, including excellent survey books on methods in bioinformatics, e.g. Sensen (2006) or systems biology, e.g. (Klipp et al. (2005). Cassman et al. (2007), in a worldwide survey of mostly individual laboratories, briefly refer to a few of the collaborative projects described in this book and give little detail on how they operate. Even so, the efficacy of these projects is recognised, e.g. "The largest of these projects create distributed virtual institutes, e.g. BIOSAPIENS, with capabilities, note American reviewers, that exceed those of comparable U.S. projects" (pp. 119–122 in Cassman et al. 2007).

Access to Resources

Researchers in principle can learn about bioinformatics and systems biology tools from the literature, from conferences, from work programmes of large institutions like the intergovernmental organisation EMBL Heidelberg (EMBL-Heidelberg 2007), the European Molecular Biology Laboratory, and the European Bioinformatics Institute (EBI 2007) of the EMBL in Hinxton. EMBL Heidelberg (EMBL-Heidelberg 2007) is a major European centre for systems biology, and the EMBL outstation EBI (2007) provides access to large numbers of databases, bioinformatics tools, services and linked information browsers. However, many researchers do not know about the European Commission projects that are linking these resources across Europe and the world and providing interconnected databases and resources and scientific networks. These projects greatly increase possibilities to link, interpret and utilise huge volumes of research data. Insights in this book into how these scientific results and tools are developed further indicate ways that individual researchers can benefit or participate: by joining future projects; by indirect and informal collaborations; by using the information and tools provided. This book shows how all the pieces are put together in broad-ranging research programmes via integrated and well-managed research programmes, strongly and regularly interacting teams from many laboratories, and the communication and linkage inherent in Internet-based research.

Research Paradigms

To say that there are new research paradigms does not mean that old ways are obsolete. The challenge is to identify where bioinformatics and systems biology methods can be applied to usefully improve knowledge and applications, and at what level of complexity. Noble (2007) argues for a "middle-out" approach, backing it with a long career of superb experiment and modelling of the heart, where analysis and experiment start at physiological levels and work in both directions. In contrast, this book reflects a diversity of pragmatic approaches, observing that successful projects use what works. Nature and evolution, via homeostatis and dominant mechanisms, have arranged that there are a large number of approaches of varying complexity that provide useful insights that support both basic and medically oriented research. Rather than the genome representing a complete "book or blueprint of life" or a universal instruction manual, it instead contains information on how to produce parts, which are then generated where and when necessary. Thus, if a part is lost or modified owing to a mutation, it is still possible to identify that some systems using that part will work differently, even if all the mechanisms are not understood. Indeed, as will be discussed for the ENCODE (2007) and GENCODE (2007) results, the chromosomal DNA sequence itself is far more complicated than just genes coding for proteins.

Complexity Levels

Researchers working as individuals still have a major role to play, when working on single genes or proteins or processes, accessing the relevant databases and tools as necessary. Similarly, loosely coupled international collaborations, such as those contributing to genome sequence databases, also have vital roles. The difference is that in Commission-funded collaborative projects, a large number of laboratories are linked by overall working strategies applied by the consortium joint management structures, with regular interchange of ideas, personnel and data, all of which within strategies they have proposed themselves. The areas of bioinformatics and systems biology are particularly apt for this collaboration, since data standardisation and exchange is almost inherent in the work style, and communication via the Internet is intrinsic to the field. This is also a contrast with looser collaborative projects, where groups of laboratories receive funding in a similar area, e.g. genome sequences, but still work separately, perhaps communicating or meeting periodically to agree on standards, protocols and priorities. Commission projects are able to evolve continuously, and to explore several approaches. Some avenues are very successful, some are interesting but blind alleys. Some unsuccessful approaches result, which are to be expected and are not penalised. Many collaborative projects have been excellent at developing solvable, state-of-the-art research areas where successful and useful models were developed.

European Commission Research Projects

Framework Programmes

A full description of European Commission supported research programmes is given in Chap. 10. Here, a very brief introduction is provided to explain the terminology used to discuss projects in the bulk of the book. The European Commission supports health research, as well as many other areas, via the Framework Programmes for Research, described on the FP5 (2007), FP6 (2007) and FP7 (2007) websites of

- CORDIS (2007), the European Community Research & Development Information Service and
- Europa (2007), the Gateway to the European Union

Research in bioinformatics and systems biology was primarily supervised by the Health Research Directorate (Health-Research 2007) by the unit for fundamental genomics (Fundamental-Genomics 2007) in FP6 (2007) and currently is primarily supervised by the unit for genomics and systems biology (Genomics-Systems-Biology 2007) unit in FP7, (see fact sheets 1 and 2). There are also major activities elsewhere in the European Commission research schemes, including Research Infrastructures, Directorate General for the Information Society and the Innovative Medicines Initiative. Full information about the entire Framework Programme is available at FP7 (2007), with a full range of supporting documents available at FP7-Find-Document (2007). Within the FP7 Specific Programme (FP7-Specific-Programme 2007), an overall European strategy for research is defined, with visions and goals. In order to carry out the strategy in the specific programme, a number of actions are taken, decided upon by a process of consultation, evaluation and assessment with the involvement of the research community and other players at all levels.

Projects

The European Commission funded several general types of projects in FP6 (FP6-instruments 2007), which allow different combinations of researchers, depending on the scale of the task. Projects are organised into work packages, and each work package requires the production of deliverables, which are often in the form of reports published on the project website. Internal project management is also central to the success of the projects. Projects are highly interactive, with scientific decisions and actions decided at appropriate levels, including project coordinator, project management and scientific boards, annual meetings of the whole project team, work package members, word package and deliverable coordinators, laboratory teams, and individual scientists. See for example the management structure of BioSapiens-Management (2007).

A Practical Guide to Project Websites, Participants, Coordinators, Publications

Websites, Participants and Coordinators

Table 1.2 lists some of the projects which are funded primarily for bioinformatics or systems biology research. These and many additional projects, which have important components of computational biology in their approach, are referred to in the following chapters. The project full name is given in the first column of Table 1.2, and in the second column, a reference is given to the project acronym and website. In the third column, a reference is listed for the coordinator's institution or research unit. The institutions participating in these projects also represent major resources with their own capabilities. The coordinating institution and the coordinator and project managers often represent the largest of these resources. On all the project websites, a full list of the participants is directly available, which are too numerous to list here directly. The coordinators are often major institutes in bioinformatics or in systems biology, with central repositories of computational resources.

Project Publications

The papers published by these collaborative research projects constitute a major resource, and provide much more information on project results. In this book, a few key publications have been referenced for each project, but the list of publications, and in turn the publications list in each publication, is far too large to reproduce here. BioSapiens alone has 75 publications listed. Most of these publication lists are available directly on the project websites, or via individual work packages or deliverables. In the fourth column of Table 1.2, the reference tab within the website referred to in column 2 is given, to facilitate access to publications lists, and to indicate if they are available.

This book does not describe detailed procedures as to how to use bioinformatics and systems biology tools and resources once they have been found, but there are a number of central institutions and their educational websites that do provide extensive information; see Table 1.3.

Table 1.2 Project acronyms, description, publications lists, partners list

Project full name	Project and participant reference	Coordinating institution	Publications list – website link
Arabidopsis growth network integrating omics technologies	AGRON-OMICS(2007)	PSB (2007)	Resource centre, then references
Systems biology of the AMP-activated protein kinase	AMPKIN (2007)	Hohmann (2007)	–
alternative transcript diversity	ATD (2007)	INSERM (2007)	Literature
Towards an understanding of dynamic transcriptional regulation at global scale in bacteria: A systems biology approach	BaSysBio (2007)	INRA (2007)	Publications
Integrative genomics and chronic disease phenotypes: modelling and simulation tools for clinicians	BIOBRIDGE (2007)	Barcelona (2007)	–
A European virtual institute for genome annotation	BioSapiens (2007)	EBI (2007)	Publications
Biosimulation – a new tool in drug development	BIOSIM (2007)	DTU-Physics (2007)	–
An integrative approach to cellular signalling and control processes: bringing computational biology to the bench	COMBIO (2007)	CRG (2007)	Groups, then publications
Computational systems biology of cell signalling	COSBICS (2007)	SBI (2007)	Publications
Dedicated integration and modelling of novel data and prior knowledge to enable systems biology	DIAMONDS (2007)	PSB (2007)	Publications
A European model for bioinformatics research and Community education – bioinformatics grid	EMBRACE (2007)	EBI (2007)	Publications
European modelling initiative combating complex diseases	EMI-CD (2007)	MPIMG (2007)	Publications
European Network of Excellence enabling systems biology	ENFIN (2007)	EBI (2007)	Publications
European Systems Biology Initiative combating complex diseases	ESBIC-D (2007)	MPIMG (2007)	Publications
Entrainment of the circadian clock	EUCLOCK (2007)	LMU (2007)	Subprojects
European fungal genomic database	Eurofungbase (2007)	LUIB (2007)	Publications
Quantifying signal transduction	QUASI (2007)	Hohmann (2007)	–
Systems biology of RNA metabolism in yeast	RIBOSYS (2007)	WTCCB (2007)	–
Streptomyces as a protein production platform	STREPTOMICS (2007)	Rega (2007)	–
Systems biology for medical applications	SYSBIOMED (2007)	DECHEMA (2007)	Workshops
The European molecular biology linked original resources	TEMBLOR (2007)	EBI (2007)	See documents in subprojects
Validated predictive dynamic model of complex intracellular pathways related to the cell death and survival	VALAPODYN (2007)	UJF (2007)	–
Yeast Systems Biology Network	YSBN (2007)	CMB-DTA (2007)	–

Table 1.3 Organisations and their educational material

Organisation	Educational material
European Bioinformatics Institute	EBI (2007) EBI–2can (2007)
US National Centre for Biotechnology Information	NCBI (2007) NCBI-Education (2007)
Centre for Information Biology and DNA Data Bank of Japan	CIB-DDBJ (2007)
MIT	MIT-Open-Courseware (2007)

Resources and Infrastructures

During the research process, extensive resources and infrastructures are being created. These are not just centralised databases and services at major hub institutions like EBI (2007), but they consist of distributed and increasingly linked resources around Europe and the world. This book provides information about these linked resources, which greatly enhance the value of the individual resources as tools. There is thus a very strong incentive for researchers to collaborate and to link their own databases, since the content then becomes much more useful than before, both to the owners of the databases and to researchers outside. Examples of these linked projects include TEMBLOR (2007) and FELICS (2007).

Accessing Knowledge and Projects

This book provides a guide to accessing the project knowledge base, via work packages, deliverables, databases and tools. In particular, with the partner list and the work package structure, it is very easy to identify which institutions and people are working in particular areas, and how their teams are formed. It is possible to do this also via the published literature and author lists, but the projects give a better insight into how the work is divided, how people collaborate, and therefore the best entry point for collaboration. The projects in this book are at very different stages: some only started recently, others are mature, such as BioSapiens (2007). There are varying proportions of plans to results in the description of each. There are many ways of collaborating with existing projects without being a contractual member. Contractual members receive project money, but also have obligations. European Commission projects often arrange large numbers of informal collaborations and information exchanges. In the following scientific areas, this book provides:

- A description of each area
- An outline of the contribution from each relevant EU collaborative project
- Pointers to key results and the resources used to generate them
- Highlights of the results from each project, often available in publications or "deliverables"
- Indications as to how to use the material and to access the collaborations

Participating in Projects

The CORDIS (2007) information service has a very extensive knowledge base about how to participate in projects in FP7 currently in progress (FP7-CORDIS 2007). There exists extensive documentation as to how to participate (FP7-Participate 2007), although the information is not yet as systematic as in FP6-step-by-step (2007). The general principles are similar: Find a relevant call at FP7-Find-a-call (2007) and review the relevant documents, especially in the health research area (FP7-Calls-Health 2007).

Key Sources

Several key sources were used to compile this book:

1. Over 50 key textbooks and over 100 major publications provide useful background and introductions to scientific areas; see References.
2. Many websites besides those of projects are included which help to demonstrate the state of the art in many fields; see References.
3. The main CORDIS (2007) website.
4. Several workshops in bioinformatics and systems biology were held to develop policy; see Table 1.1.
5. Details of the projects can be found on their websites quoted in the reference section of this book, and at CORDIS-Projects (2007) using the search box and acronym.
6. Many projects are discussed in detail in a series of newsletters at Fundamental-Genomics (2007), some of which are older, some of which are up to date.
7. Summaries of European Commission projects, and introductions to the various areas, are found in project summary books of areas of health and related life sciences research; see:

   – Ingemansson and Knezevic (2005) – biotechnology
   – Manoussaki (2006) – cancer
   – eHealth (2007)
   – Food (2007)
   – Kyriakopoulou et al. (2007) – genomics
   – Ghalouci (2007) – infectious diseases
   – Vanvossel (2005) – major diseases
   – Schmaltz (2007) – pneumonia
   – Research-Infrastructures (2007)
   – Joliff-Botrel and Perrin (2007) – stem cells

Part I
Fundamental Collaborative Research and Computational Biology

Chapter 2
Bioinformatics

Abstract Bioinformatics is a major research area in its own right, as well as a source of tools, databases and services. This research aspect is highlighted in the area of genome annotation, in its broadest sense of defining the biological role of a molecule in all its complexity. This complexity is explored in this chapter, and involves gene definition, alternative transcripts and splicing, gene regulation and expression, the functional annotation of proteins, post-translation modification, membrane and localisation prediction, protein complexes, networks and pathways. Annotation is further unified in an international collaborative effort on compiling an encyclopaedia of DNA elements.

Introduction

Genome Sequences

The genome projects have revealed and codified the entire DNA sequence of humans and other organisms, which if not entirely providing a "blueprint" of life describes many key elements. The first draft of the human sequence was published in 2001 (The International Human Genome Mapping Consortium 2001), and there are now over 53 eukaryota, 46 archaea and 517 bacteria genome sequences in the public domain (EBI 2007). This explosion in genomic information has been achieved in a remarkably short period of time, and the flood of new sequence data is likely to continue for the foreseeable future. However, a DNA sequence is a string of letters; it must be interpreted in terms of the RNA and proteins that it encodes and the promoter and regulatory regions that control transcription and translation.

Annotation

Annotation can be described as the process of "defining the biological role of a molecule in all its complexity" and mapping this knowledge onto the relevant gene

F.B. Marcus, *Bioinformatics and Systems Biology*,
DOI 10.1007/978-3-540-78353-4,

products encoded by genomes. This involves both experimental and computational approaches and, indeed, absolutely requires their integration.

European Contributions

European scientists have been very active in the field of genome and protein annotation, with Ensembl (2007) and Swiss-Prot (2007), now integrated in UniProt (2007), being among the primary resources in use worldwide. Many of the tools used in genome and protein sequence and structure annotation, prediction and validation, as well as in pathway analysis and secondary resources derived from protein sequences and structures were developed in Europe. The fragmentation of resources for genome annotation meant that only a few bioinformatics experts knew where to look for them. Consequently, most experimentalists had been unable to access all the best information about a genome. In what follows, key recent contributions from several projects are discussed, including ATD (2007), BaSysBio (2007), BioSapiens (2007), BioBabel (2007), ENCODE (2007), EuTRACC (2007), GO (2007), IIMS (2007), SPINE (2007) and TEMBLOR (2007),

Key Areas

Some of the key research areas in the bioinformatics of genome annotation are systematically discussed in this chapter, and include:

- Gene definition/alternative splicing
- Regulators and promoters
- Expression
- Genetic variation (haplotypes, single-nucleotide polymorphisms, etc.)
- Protein families, orthologues
- Membrane proteins and ligands
- Three-dimensional protein structure
- Post-translation modification and localisation
- Sequence and structure to function
- Protein–protein complexes
- Pathways and networks

Genome Annotation

A European Virtual Institute for Annotation

In response to the topic published in support of genome annotation (FP6-2002-LIFESCIHEALTH, 2002), the BioSapiens (2007) project was successfully proposed, and has created a fully functioning European Virtual Institute for Genome

Annotation BioSapiens (2005). This virtual institute has established tools and work flows that allow annotation over a large part of the range of biological knowledge, and it addresses the full range of research topics listed above. The institute is improving bioinformatics research in Europe, by providing a focus for annotation and through the organisation of European meetings and workshops to encourage cooperation, rather than duplication of effort.

Distributed Annotation

An important aspect of the network activities is to achieve closer integration between experimentalists and bioinformaticians, through a directed programme of genome analysis, focused on specific biological problems. The annotations generated by the Institute and external participants are available in the public domain and are easily accessible on the Web. This has been achieved through a distributed annotation system (DAS 2007), which has evolved to take advantage of the GEANT2 (2007) pan-European research and education network supported by Enabling Grids for E-Science in Europe (EGEE 2007). The groups also focus on the development of improved computational methods for annotation through new methods available via the Web. Annotations from these new methods are available via DAS (2007), which is available via the website as a DAS portal and has a DAS server information service (DAS-Information 2007). There are over 23 distinct DAS servers providing 69 different data sources.

An Integrated Approach

Many of the tools used in genome and protein sequence and structure annotation, prediction and validation, and pathway analysis have been developed in Europe. BioSapiens (2007) has been instrumental in creating increased integration, expert training and improved tools and services, and an enhanced European role in the academic and industrial exploitation of genomics. Some of the main results being produced by the project include the development of an integrated approach to genome annotation from gene to function, and ultimately the establishment of an integrated and distributed website for genome annotation. A description of the individual research areas is available via the work packages as shown in Table 2.1.

Annotation Deliverables

References to the bioinformatics methods used in each particular area are available via deliverables within the project work package descriptions. Work packages in the range of more than 100 are for work in progress. For example, in the area

Table 2.1 BioSapiens scientific areas for genome annotation

Scientific area	Reference
Gene definition/alternative splicing	BioSapiens-WP1 (2007) BioSapiens-WP101 (2007)
Gene regulation and expression	BioSapiens-WP2 (2007) BioSapiens-WP3 (2007) BioSapiens-WP102 (2007)
Variation (haplotypes and single-nucleotide polymorphisms)	BioSapiens-WP4 (2007) BioSapiens-WP103/110 (2007)
Functional annotation of proteins	BioSapiens-WP5 (2007) BioSapiens-WP7 (2007) BioSapiens-WP9 (2007) BioSapiens-WP104 (2007)
Post-translational modification, membrane and localisation prediction	BioSapiens-WP6 (2007) BioSapiens-WP8 (2007) BioSapiens-WP105 (2007)
Protein complexes, networks and pathways	BioSapiens-WP10 (2007) BioSapiens-WP11 (2007) BioSapiens-WP106 (2007)

of gene definition/alternative splicing (BioSapiens-WP101 2007), deliverable "Del 1.1: A list of experimentally validated gene structures for the human genome and other mammalian genomes" contains a full list of references at the end of the document, including a reference to the major tool used, Ensembl (Hubbard et al. 2005), supplementing the overall project list of more than 75 major publications at BioSapiens-Publications (2007)

Genome Browser and Distributed Annotation Viewer

Ensembl (2007) is a software system which produces and maintains automatic annotation on selected eukaryotic genomes. Ensembl provides accurate and automatic analysis and annotation of genome data, concentrating on vertebrate genomes, but includes a wide range of other model organisms. There are over 33 genomes available in Ensembl; these include those of human, several other mammals, chicken, four species of fish, several insect species and a nematode. Prereleases of three further organisms, including the start of a large number of low-coverage mammalian genomes such as that of elephant, are available through Pre-Ensembl. Ensembl automatically annotates genome sequence and predicts the positions of genes, to provide a comprehensive range of sequence features and genome-wide gene and protein sets. The system is applied in a consistent way to different species, and incorporates between-species comparisons of genome sequence and homologous genes. A rich variety of links to external databases helps to make Ensembl a key starting and central reference point for studies in genetics and molecular biology. Ensembl continues to improve annotation to both the human and the mouse genomes. It also provides timely annotation, both for newly sequenced genomes (such as that of platypus) and for previously sequenced genomes for which information continues to be refined (such as that of chicken). In particular, Ensembl has an effective pipeline for calculating non-coding RNA gene models. These include structural RNAs, such as U6 RNA, and regulatory RNAs, such as micro-RNAs. In the human genome there are also results on *cis*-regulatory networks. This is an actively growing area of research with many groups developing methods

As a distributed annotation viewer, the Ensembl (2007) genome browser is used with its DAS (2007) display to look at the p53 gene, an apoptosis-regulating gene with nearly 44,000 publications, shown in Fig. 2.1, providing the following information, and much more:

Gene, TP53 (HUGO Gene Nomenclature Committee, HGNC, symbol); synonyms, p53.
This gene is a member of the human CCDS set: CCDS11118.
Ensembl gene ID: ENSG00000141510
Genome location: This gene can be found on chromosome 17 at location 7,512,464–7,531,642.
The start of this gene is located in contig AC087388.9.1.121017.

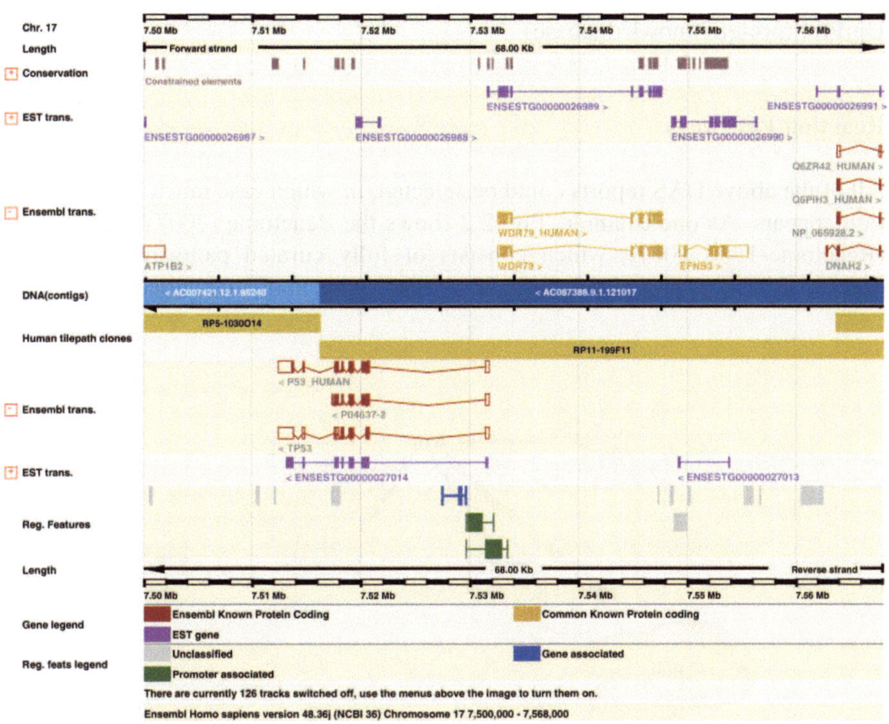

Fig. 2.1 Ensembl (2007) browser with DAS display to look at p53, Ensembl gene ID ENSG00000141510

Description: Cellular tumour antigen p53 (tumour suppressor p53) (phosphoprotein p53) (antigen NY-CO-13). Source, UniProt/Swiss-Prot P04637
Prediction method: Genes were annotated by the Ensembl automatic analysis pipeline using either a GeneWise/Exonerate model from a database protein or a set of aligned complementary DNAs (cDNAs) followed by an open reading frame prediction. GeneWise/Exonerate models are further combined with available aligned cDNAs to annotate untranslated regions (for more information see Curwen et al. 2004).
Gene DAS report: DAS sources
AltSplice (alternative Splice database)
AltTrans (alternative transcript diversity database)
ArrayExpress (gene expression database)
GAD (genetic association database)
HGNC (HUGO Gene Nomenclature Committee)
HUGO_text (PubMed text mining via HGNC symbol)
Phenotypes (associated directly or via orthologues or protein families)
Protonet (global classification of proteins into hierarchical clusters)
RZPD verif. cDNA (RZPD sequence verified non-redundant cDNA clone sets)
RZPD esiRNA (RZPD gene silencing (RNA interference) resources)
RZPD Prot Exp (RZPD clones ready for protein expression)
Reactome (knowledgebase of biological processes)
UniProt (protein knowledgebase)

Reaction Pathways

All of the above DAS reports could be selected, in which case much more information appears. As one example, Fig. 2.2 shows the Reactome (2007) path provided (Reactome-1756 2007), which consists of fully curated pathway data, and is described as phosphorylation of p53 at ser-15 by ataxia-telangiectasia mutated

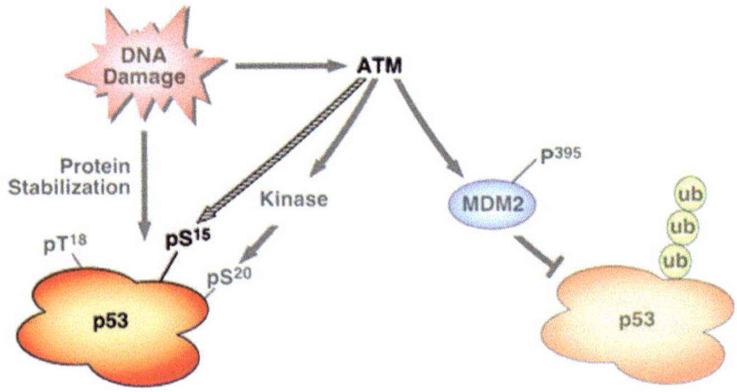

Fig. 2.2 Reactome-1756 (2007), A curated knowledgebase of biological pathways – path 1756 – phosphorylation of p53 at ser-15 by ataxia-telangiectasia mutated (*ATM*) kinase (*Homo sapiens*)

(ATM) kinase stable identifier REACT_1756.1. In response to DNA damage due to ionising radiation, the serine at position 15 of the p53 tumour suppressor protein is rapidly phosphorylated by the ATM kinase. This serves to stabilise the p53 protein. A rise in the levels of the p53 protein induces the expression of the p21 cyclin-dependent kinase inhibitor. This prevents the normal progression from G1 to S phase, thus providing a check on replication of damaged DNA.

Experimental–Computational Collaboration

BioSapiens (2007) also stimulates cooperation between experimental scientists and computational biologists for genome annotation, in the form of meetings and joint collaborations. Experimental validation of predictions made *in silico* forms part of these collaborations. The tools are also validated and applied via thematic work packages.

Thematic Collaborations

BioSapiens (2007) consciously chooses particular thematic areas where the full power of the Virtual Institute can be directed towards particular scientific problems. These areas are summarised in Table 2.2. These thematic areas show the power of these large networks, since they can apply the tools they develop to a wide range of problems, including relevant disease research. The disease themes are discussed later in the appropriate book sections. The exploitation of the biological information enabled by BioSapiens (2007) will in some cases be relatively direct, e.g. improved health-care through better drugs, new vaccines and personalised medicines for individuals and subpopulations, and improved understanding of diet and health.

Critical Mass of Resources

BioSapiens (2007) has had an important impact on the establishment of a European research structure that supports the coordination of bioinformatics research activities across different subareas, and across different areas of medical and biotechnological application. It has developed the required level of critical mass so that Europe, with primarily nationally based funding schemes, can compete with the major investments

Table 2.2 BioSapiens (2007) thematic areas of scientific collaboration

Infectious diseases	BioSapiens-WP15 (2007)
Down syndrome	BioSapiens-WP16 (2007)
ENCODE project	BioSapiens-WP20 (2007)
	BioSapiens-WP108 (2007)
Cancer	BioSapiens-WP109 (2007)

made in the USA and Japan. The integration between the groups has already had a lasting impact on the European bioinformatics infrastructure, and on the sharing of human resources, infrastructure databases and tools. Through cutting-edge research, high-level training and vigorous European-level interaction, BioSapiens has made a substantial contribution to improving Europe's knowledge base.

Bioinformatics Tools For Annotation

Integrated Tool Development

The TEMBLOR (2007) project, the European molecular biology linked original resources, received almost €20 million over 3 years. The project concentrated on research and development to build major bioinformatics resources. These resources were embedded in an integrated layer known as Integr8 (2007), allowing biomedical researchers to fully exploit genomic and proteomic data. Integr8 draws on databases that are maintained at major bioinformatics centres in Europe, and also on important new resources. The main aim of TEMBLOR (2007) is to allow users to carry out complex queries across databases in a much simpler way than has previously been possible, by accessing all of these databases through Integr8.

A summary of the projects within TEMBLOR includes:

- Integr8 (2007) – an integrated layer for the exploitation of genomic and proteomic data
- EMSD (2007) – storing and analysing the structures of large molecules
- DESPRAD (2007) – standards and repositories for gene expression experiments
- IntAct (2007) – standards and resources for protein–protein interaction data

Integrated Layer for Genomic and Proteomic Data

Integr8 (2007), described by Kersey et al. (2005) is a Web portal for exploring the biology of organisms with completely deciphered genomes. For 53 eukaryota, 46 archaea and 517 bacteria, Integr8 provides access to general information, recent publications, and a detailed statistical overview of the genome and proteome of the organism. Integr8 (2007) also provides access to complete genomes and proteomes, as part of developing integrated search capabilities, resulting in a major strengthening of individual database capabilities for protein sequence work, taxonomy and ontologies, via support given to projects such as UniProt (2007), InterPro (2007), NEWT (2007), GO (2007) and GOA (2007), which had already been partially developed by BioBabel (2007). Although these databases are centralised at the EBI, their establishment and continuing development are based on multilaboratory collaboration in the development of Integr8 (2007) and Genome Reviews (2007). The Integr8 (2007) Web portal provides easy access to integrated information about deciphered genomes and their corresponding proteomes. Available data include:

- DNA sequences from databases including the EMBL nucleotide sequence database, Genome Reviews, and Ensembl
- Taxonomy of the organism via NEWT (2007)
- Protein sequences from databases including the UniProt knowledgebase and IPI (2007)
- Statistical genome and proteome analysis performed using InterPro (2007), CluSTr (2007), and GOA (2007)
- Information about orthology, paralogy, and synteny

Protein Structure

The Macromolecular Structure Database (MSD 2007) group is one of the three partners in the worldwide Protein Data Bank (PDB), the consortium entrusted with the collation, maintenance and distribution of the global repository of macromolecular structure data, especially protein structure data. The PDB is the international repository for three-dimensional structures of macromolecular complexes of proteins, nucleic acids and other biological molecules. The data range from those of small protein fragments to those of large macromolecular assemblies such as viruses and ribosomes, whose structures have been determined by experimental methods such as X-ray crystallography, nuclear magnetic resonance spectroscopy or electron microscopy. Many of the electron microscopy analysis capabilities were developed in the IIMS (2007) project. These data are publicly accessible, and are used by scientists, researchers, bioinformaticians, educators, students and lay audiences. By annotating and archiving the data in an efficient and consistent way, the PDB supports the understanding of biological phenomena at a structural level and facilitates new discoveries in science. The MSD (2007) tools available include:

- MSDlite (simple search of relational PDB
- MSDpro (advanced search system)
- MSDmotif (small three-dimensional motif statistics with extensive Φ, Ψ, χ search options)
- MSDtemplate (local residue interactions in the PDB
- MSDpisa (search and analysis of protein interfaces, surfaces and assemblies)
- MSDchem (ligand search)
- MSDmine (ad hoc queries and data analysis)
- MSDsite (ligand-environment search)
- MSDfold (secondary structure matching)
- MSDanalysis (validation and analysis of MSD data)
- MSDtarget (sequence target search)
- EMsearch (search the electron microscopy database)
- MSDbar (search system using toolbar application)

- PQS (protein quaternary structure server)
- PQS-Quick (simple PQS search)
- NMR Representatives (representative model from NMR ensemble)
- Reference Server (search by author/ID for PDB structures without final reference)
- Relibase (a program for searching protein-ligand databases)
- Biotech (validation suite for protein structures)
- Search OCA (enter OCA search system)
- PDB Pending (search pending and waiting list for status of file under processing)
- PDB New Entries (PDB latest releases)
- SPINE @ EBI (direct to spine targets)

The tools of MSD (2007) provide a wide range of options, for example visualising protein structures that were analysed in the FP5 (2007) collaborative project SPINE (2007) – Structural Proteomics in Europe. Figure 2.3 shows an example of a surface protein of a cancerous cell:

Expression Data

The DESPRAD (2007) subproject was aimed at developing ArrayExpress (2007), a public repository for microarray data, and the standards and ontologies needed to describe, exchange and store microarray data (experiments, protocols and array designs). Also, software tools were developed for querying the database, and for

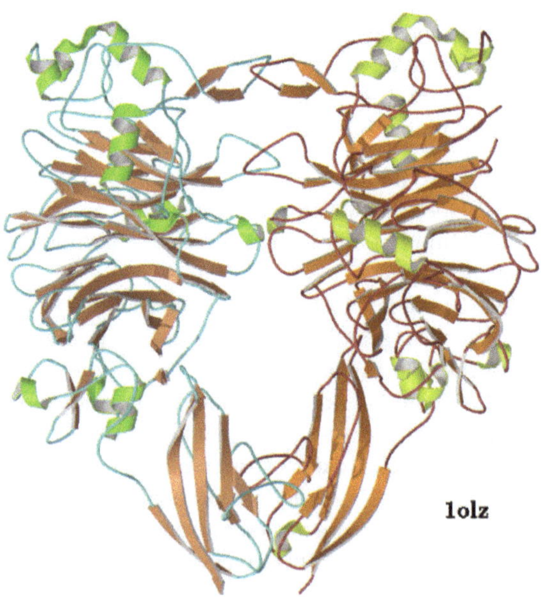

1olz

Fig. 2.3 Semaphorin 4D precursor protein structure, *Homo sapiens*, Protein Data Bank accession 1OLZ, Swiss-Prot accession Q92854 (see also Love et al. 2003)

curation and submission of data. Analysis tools, standalone or integrated with the database, were also goals for this project.

Elements of this project include:

- Minimum Information About a Microarray Experiment (MIAME 2007). This has become an accepted worldwide standard.
- Microarray and Gene Expression (MAGE 2007). These standards have been adopted by The Object Management Group (OMG 2007).
- MGED (2007) is an ontology for describing microarray experiments.
- ArrayExpress (2007) is a public repository which is online and accepting submissions.
- MIAMExpress (2007) is a MIAME (2007) compliant microarray data submission tool.
- Expression-Profiler (2007) is an open, extensible Web-based collaborative platform for microarray gene expression, sequence and protein–protein interaction data analysis.

Storing and Interpreting Microarray Data

ArrayExpress (2007) is now one of the major tools of modern biology, essential for storing and interpreting microarray data. ArrayExpress is a public repository for microarray data, which is aimed at storing MIAME-compliant data in accordance with MGED (2007) recommendations. The ArrayExpress (2007) data warehouse stores gene-indexed expression profiles from a curated subset of experiments in the repository. Public data are made available for browsing and querying on experiment properties, submitter, species, etc. Queries return summaries of experiments and complete data, or subsets can be retrieved. A subset of the public data are reannotated to update the array design annotation and curated for consistency. These data are stored in the data warehouse and can be queried on gene, sample, and experiment attributes. The results return graphed gene expression profiles, one graph per experiment.

Microarray Expression, Sequence and Protein–Protein Data Analysis

Coupled to ArrayExpress is Expression-Profiler (2007). Expression Profiler: Next Generation (Kapushesky et al. 2004) is an open, extensible Web-based collaborative platform for microarray gene expression, sequence and protein–protein interaction data analysis, exposing distinct chainable components for clustering, pattern discovery, statistics (via the R programming language), machine-learning algorithms and visualisation. The architecture modularises the original design and allows individual analysis-task-related components to be developed by different groups and yet still seamlessly to work together and share the same user-interface look and feel. Data analysis components for gene expression data before processing, missing value imputation, filtering, clustering methods, visualisation, significant gene findings, between-group analysis and other statistical components are available from the EBI (2007) website. The Web-based design of Expression-Profiler (2007) supports data sharing and collaborative analysis in a secure environment. Developed tools are integrated with the microarray gene expression database ArrayExpress (2007) and form the exploratory analytical front end to those data.

Protein–Protein Interactions

Major capabilities for studying protein–protein interactions were established by the TEMBLOR (2007) subproject IntAct (2007). IntAct (2007) provides a freely available, open source database system and analysis tools for protein interaction data. All interactions are derived from literature curation or direct user submissions and are freely available. An experiment consists of many interactions which contain two or more interactors. An interactor can be either an individual protein or a protein complex (i.e. the result of a previous interaction). Therefore, an interaction can consist of two or more proteins or complexes. Each object (experiment, interaction, interactor) has attributes assigned which provide a detailed description. This is important as specific features of an experiment can have a profound impact on the type of interaction. Every IntAct (2007) object has a unique accession code which starts with "EBI-", followed by a number. It is these accession codes that enable the hierarchical data structure.

Protein Sequence and Function Database

UniProt (2007) is the world's most comprehensive catalogue of information on proteins. It is a central repository of protein sequence and function created by joining the information contained in UniProt Knowledgebase (UniProtKB)/Swiss-Prot, UniProtKB/TrEMBL, and PIR. UniProt is composed of three components, each optimised for different uses. UniProtKB is the central access point for extensive curated protein information, including function, classification, and cross-reference. The UniProt Reference Clusters (UniRef) databases combine closely related sequences into a single record to speed searches. The UniProt Archive (UniParc) is a comprehensive repository, reflecting the history of all protein sequences.

Protein Sequence Grouping

InterPro (2007) is a searchable database providing information on sequence, function and annotation. Sequences are grouped on the basis of protein signatures or "methods". These groups represent superfamilies, families or subfamilies of sequences. The groups may be defined as families, domains, repeats or sites. The function of sequences within any group may be confined to a single biological process or it may be a diverse range of functions (as in a superfamily) or the group may be functionally uncharacterised, but without exception every entry has an abstract and references are provided where possible. It is well worth browsing the database and going through the InterPro frequently asked questions.

Gene Definition/Alternative Transcripts and Splicing

Gene Definition

Although the human genome sequence has been available in at least draft form for several years, the complete list of all of its functional regions is far from complete (BioSapiens-WP1 2007). Genome annotation relies on computational methods to integrate information from both de novo gene prediction algorithms and protein databases and other sources of expressed sequences such as expressed sequence tags (cDNA) and high-quality reference sequence messenger RNAs (mRNAs). Each of the sources of expressed sequence must be accurately mapped to the exact genome locations corresponding to the sequence to discover the gene responsible for the given sequence. These processes require significant computational resources. There is still considerable question about the total number of protein-coding genes contained within the human genome. The currently accepted estimate is approximately 25,000 genes, and many may include multiple transcribed forms. This estimate is based on a number of independent methods for annotating the human genome (e.g. Curwen et al. 2004). Essentially the same number of genes are thought to be present in the mouse genome.

Alternative Transcripts and Splicing

A single human gene can produce a variety of alternative transcripts (or mRNA isoforms) (pp. 436–437 in Alberts et al. 2002), which differ in terms of their transcription initiation, splicing or polyadenylation patterns. Expression of alternative transcripts has been observed to be specific to tissue type or developmental stage. Disruptions in alternative transcript expression have serious consequences for an organism and are associated with numerous diseases, including cancer, multiple sclerosis, heart failure and neurodegenerative disorders.

Gene Definition and Alternative Splicing Methods

In BioSapiens (2007), the goals in the areas of gene definition and alternative splicing (BioSapiens-WP1 2007; BioSapiens-WP101 2007) are to study functional regions of genomes, in particular the genes, focusing on four main areas of investigation:

1. The basic gene structure (intron–exon structure)
2. The presence of differential gene structure (alternative splicing)
3. The evolution of gene structure
4. The alternative splicing process

Methods involve combining classical machine learning algorithms, theoretical studies of evolution and experimental techniques, and using genomes across all eukaryota where appropriate, but with a focus on mammalian and in particular human, mouse and rat genomes. As well as multigroup interactions, providing the crucial feedback loop between experiments and predictions, there is networking with outside groups; in particular, those with prediction algorithms that influence the understanding of functional gene content (e.g. signal peptide prediction and structural modelling of protein sequences), and with the thematic disease foci by providing in-depth analysis of genomic regions and genes of interest.

Alternative Transcription Goals

The Alternate Transcript Diversity (ATD 2007) project has investigated the mechanisms responsible for the formation of different alternative transcripts. These mechanisms are discussed extensively in the ATD (2007) project "literature" section, containing general references and project publications. All public deliverables are available on the website under "ATD data releases". It is also anticipated that studies in the field of alternative transcripts will have direct applications for pharmaceutical industries. These applications include disease diagnosis or prognosis of risk patients, as well as identification of new drug targets. ATD (2007) is a follow-on project from the Alternate Splicing Diversity project (Stamm et al. 2006; Thanaraj et al. 2004).

Alternative Transcription Methods

ATD (2007) is a collaborative multidisciplinary project. It has comprehensively characterised alternative transcript forms throughout the human genome, and has assessed the differential expression of these forms in time and space, in normal and disease-related tissues. This was accompanied by quality control procedures, such as research for evolutionary proof through comparative sequence data analysis, between human and mouse. Further characterisation of alternative transcripts was implemented through activities such as identification of regulatory patterns, and derivation of expression states (i.e. expression specificity in terms of association with diseases, developmental stages, or tissue specificity). The project developed standard vocabularies and models that represent gene structures and their expression patterns. The validity of the bioinformatics prediction of disease-specific alternative transcripts has been examined through the execution of reverse transcription polymerase chain reaction experiments on selected tissues. The discovery effort was accompanied by database integration, and also by dissemination to the scientific community.

Alternative Transcription Results

Some major results of ATD (2007) are summarised in two publications (Le Texier et al. 2006; Stamm et al. 2006), as follows:

- The creation of a unified ATD (2007) database integrating various information levels such as gene, feature variants, transcript variants, annotations, derived expression states, protein functionalities, results of experimental validations and associations with diseases. Fully developed query interfaces and toolboxes were available in the databases created by ASD (2007) and ATD (2007), which have been combined and upgraded to create the ASTD (2007) database.
- The definition of standards to represent gene structures and variants, and the creation of vocabularies for the representation of annotations.
- The confirmation of differentially expressed alternative transcripts in healthy and diseased tissues from human and mouse.
- The prediction of the regulatory motifs involved in alternative transcript formation.

Future Research

Traditional molecular biology approaches founded on a "one gene at a time" basis are no longer practical when detecting new disease-specific alternative transcripts. There is currently a need for the execution of genome-wide alternative transcript detection, followed by high-throughput analysis of transcript expression.

Gene Regulation and Expression

Gene Regulation and Expression Processes

Gene expression is the process by which the DNA sequence is transcribed into a gene product such as a protein or RNA. Several steps in the gene expression process may be modulated, including the transcription step and the post-translational modification of a protein. Gene transcription regulation gives the cell control over structure and function, and is the basis for cellular differentiation, morphogenesis and the versatility and adaptability of any organism.

DNA Microarray Data

A DNA microarray is a collection of microscopic DNA spots, commonly representing single genes, arrayed on a solid surface. DNA arrays are commonly used for expression profiling, i.e. monitoring expression levels of thousands of genes simultaneously. Software to store and analyse this type of data was developed in a subproject within a major European Commission infrastructures collaborative project (TEMBLOR 2007) with four major components, one of which was DESPRAD (2007), which resulted in the development of ArrayExpress (2007).

Expression Research Goals

The BioSapiens (2007) goals for gene expression (BioSapiens-WP3 2007) have been:

- Development of methods and tools enabling the building of a human gene expression compendium characterising expression patterns of all genes in different tissues and cell types in different states by integration and analysis of data from a variety of sources, including testing and application of these methods
- Development and testing of methods for using gene expression and comparative genomics data for promoter prediction and analysis
- Development and evaluation of statistical and algorithmic methods for the analysis of gene expression data in the context of biological networks

This work addresses some of the major questions in modern biology. BioSapiens-WP3 (2007) deliverable DE3.4, "Documentation on DAS links to ArrayExpress (2007) [Au3] Data", describes how a human gene expression compendium has been constructed by developing a protocol that links human gene sequences and their annotations with expression profiles. A major part of the implementation of this protocol has been the creation of a method that parses files describing the design elements comprising microarrays, with the presence of links to DNA, protein or model organism databases.

Expression Results

Several types of analysis have been performed, an example being given in BioSapiens-WP3 (2007) deliverable 3.2, "Documentation on comparative analysis of mammalian gene expression". DNA microarrays were used to characterise gene expression patterns in skin biopsies from individuals with a diagnosis of systemic sclerosis with diffuse scleroderma, and these patterns were compared with those of gene expression seen in biopsies from normal unaffected individuals. The expression profiles of the transcription factor genes, and genes exhibiting correlated expression, were obtained from these microarray experiments (stored in the data warehouse) that compared transcription patterns in organism parts and disease states. These analyses were designed to determine whether orthologous transcription factors control the expression of the same sets of genes, and in the same tissues, in both humans and mice. Additionally, the determination of the patterns of expression of human-specific transcription factors highlights divergences in basic biological processes between the two species. Furthermore, the identification of the misregulation of transcription factors in the various disease states may give a greater understanding of the molecular mechanisms underlying a disease.

Gene Regulation Research Goals

In BioSapiens (2007) gene regulation and expression studies (BioSapiens-WP2 2007; BioSapiens-WP3 2007; BioSapiens-WP102 2007), the main goals for regulation research are:

- Implementation and further development of novel sequence annotation tools for promoter analysis in the human genome
- Development of statistical methods and tools enabling the prediction of likely regulators for given genes or groups of genes
- Discovery of *cis*-regulatory modules in mammals, and the development and analysis of gene regulatory network models utilising gene expression data
- Development and testing of a similarity search engine for expression data repositories
- Improvement of predictive methods for higher organisms and interoperability of tools

Regulation Results

This work has already led to a number of tools and results, for example as reported in BioSapiens-WP2 (2007) deliverable De2.7, "Report on the utility of Web services for *cis*-regulation". They are in the process of linking worldwide databases from the following sources:

- RSAT (2007)
- Ensembl (2007)
- T-Reg (2007)
- ArrayExpress (2007)
- STRING (2007)
- RegulonDB (2007)
- ORegAnno (2007)

Several important results have been obtained, for example those reported in BioSapiens-WP2 (2007), in De2.6, report on the analysis of multiple ChIP-chip (chromatin immunoprecipitation on a DNA microarray chip) datasets in human. Results from the multiple ChIP-chip datasets produced within the ENCODE (2007) consortium are leading to an expanded understanding of the complexity of mammalian transcription. These experiments have defined approximately 4,500 transcription start sites within the ENCODE regions of the human genome. This is approximately 10 times the number of known genes in these regions and highlights the increasing complexity of mammalian transcription as described within the ENCODE project. This analysis of ChIP-chip data from multiple experiments has also demonstrated that the binding of transcription factors is symmetric around the transcription start site. This result will affect other assays for promoter regions that have traditionally concentrated only on the regions immediately upstream of the transcription start site. A key publication discussing the ENCODE (2007) and GENCODE (2007) work is Koch et al. (2007).

Systems Biology of Transcription and Regulation

Two new projects have begun, (see BaSysBio (2007) described in Chap. 4), which are taking a full systems biology approach and which will be making major contributions to data generation concerning transcription regulation and expression. BaSysBio (2007) has the goal of understanding of dynamic transcriptional regulation at global scale in bacteria. The European Transcriptome, Regulome and Cellular Commitment Consortium (EuTRACC 2007), will participate in the International Regulome Consortium (IRC 2007).

Functional Annotation of Proteins

Protein Sequence, Structure and Function Integration

Much of the bioinformatics structure for analysis of protein sequence, structure and annotation was established, extended or improved in the TEMBLOR (2007) project, via the European Molecular Structure Database (EMSD 2007),, for protein structure, and Integr8 (2007) for integration of protein sequence and structure related data. EMSD (2007) provided the basis for the current MSD (2007), the EBI Macromolecular Structure Database, which is the European project for the collection, management and distribution of data about macromolecular structures, derived in part from the PDB (2007). A wide range of structure and sequence databases have been developed, along with many tools to infer protein sequence from gene sequence, and protein structure from protein sequence. A major effort has also been under way to comprehensively link these resources. Much of this effort has occurred in European collaborative programmes, as follows.

Sequence to Structure to Function Results

BioSapiens (2007) has built on this major infrastructure of databases and tools to mount a major annotation programme for proteins (BioSapiens-WP5 2007; BioSapiens-WP7 2007; BioSapiens-WP9 2007; BioSapiens-WP104 2007). This work has been focused on the integration of methods for the construction and the validation of three-dimensional models of protein structures. The models incorporate confidence values both at the protein and at the residue level and internal quality checks. These predictions, based on the family analysis and integrated with the predictions of integral membrane proteins, were channelled to the other participants for structure-based functional annotation. Functional information gained from structure analysis is highly complementary to that obtained by high-throughput experimental, sequence-based, or genomic context methods. Despite the fact that the relationship between structure and function is a central problem in molecular biology, and thus is critical for protein engineering and drug design, there are only a few methods able to generate function predictions from the analysis of protein structures. As a consequence, many proteins with known structure are not yet functionally characterised. Methods are being established for fully automatic inference of structure from function, aiming at the identification and characterisation of functional regions in proteins. An integrated Web resource was implemented at EBI (2007). Contributions were provided by the connections to the Web server implementations of their corresponding methods by the participating groups. The results involve:

- Combining very different methods into a working pipeline
- Comparing and benchmarking the predictions to obtain a combined approach
- Contributing to the annotation of binding (and specificity) sites in protein models

Functional Sites Results

An example result for determining functional sites is found in BioSapiens-WP9 (2007) deliverable De9.14, “A Web tool for the prediction of residues of functional specificity from multiple alignments”. TreeDet (2007) (Carro et al. 2006) predicts evolutionary importance and functional sites in protein families. The server integrates the results of three separate methods for the prediction of residues of functional interest in protein families. These tree-determinant methods are based on the relation between sequence conservation and evolutionary importance and include a tree-based method, a correlation-based method and a method that employs a principal component analyses coupled to a cluster algorithm. Accurate alignments are crucial to the prediction of tree-determinant residues and for that reason a tool for the evaluation of alignment reliability (SQUARE) has been included in the package.

Small-Ligand Binding

Some of the major problems related to metabolism and drug development are related to the binding of small ligands to proteins. An example of the important results being obtained in this area is found in BioSapiens-WP9 (2007) deliverable 9.9, "New methods for characterising ligand binding sites". Stockwell and Thornton (2006) observed that the phenomenon of molecular recognition, which underpins almost all biological processes, is dynamic, complex and subtle. They presented an analysis of the conformational variability exhibited by three of the most ubiquitous biological ligands in nature, ATP, NAD and FAD, and demonstrated qualitatively that these ligands bind to proteins in widely varying conformations, including several cases in which parts of the molecule assume energetically unfavourable orientations. Several other results are presented that are fundamental to structure to function interpretations concerning proteins and ligands.

Future Plans

BioSapiens's (2007) plans for the future aim at establishing methods for fully automatic inference of protein function. The main goal will be the identification and characterisation of functional regions in proteins. An integrated Web resource will be provided through the BioSapiens (2007) portal, through the DAS (2007) protocol or by Web services. The objectives are:

- To develop tools to improve the classification of proteins, using sequence and structure information, into protein domain familie
- To improve methods for modelling protein structures from sequence and to develop quality indicators for different structures.
- To develop new methods for functional annotation from sequence and structure
- To make all the knowledge generated available through the BioSapiens (2007) portal utilising DAS (2007) where appropriate or Web services

Post-translation Modification, Membrane and Localisation Prediction

Membrane Proteins and Results

An important basis for the work of BioSapiens-WP6 (2007) on membrane proteins was established by annotation of integral membrane proteins, in terms of function, subcellular localisation and topology/structure. New methods were developed that used experimental and theoretical results from widely different studies to enhance the transmembrane topology prediction. New molecular-class specific information systems were created to better present the results to bioscientists. An example of the results is found in BioSapiens-WP6 (2007) deliverable De6.9, "Integration of the Ensembl 2.0 predictor into the TRAMPLE (2007) transnsmembrane protein labelling environment", which describes the results of joint work on the task of annotating *in silico* all the available sequences of the human genome according to the UniProt (2007) database previously selected. A new integrated and browsable database was developed that improves the previously developed TRAMPLE (2007). The new environment/Web server/DAS (2007) server is called PONGO (2007), a Web server for multiple predictions of all-α transmembrane proteins (Amico et al. 2006). It is based on a relational database containing all the data generated. The local DAS (2007) annotation server is resident on the same machine. The results of the effort involve the inclusion of new methods for the *in silico* annotations of transmembrane predicted regions. The user is able to trace for each UniProt (2007) sequence whether the protein is or is not endowed with a signal peptide, whether the sequence is or is not a membrane protein, and its putative topology, as computed by six different predictors, two of which are newly included in the latest version of the Web server. This allows users to compare among different predictors at the same time and assess whether the expected results are in agreement with their own experimental findings. Alternatively, different predictions, especially when in agreement, may enforce the expectation that a given chain is a membrane protein and in this case the putative topology may help in designing experiments in order to validate (or not) the number of transmembrane helices and the location of the N and C termini of the protein with respect to the plane of the membrane. This may be particularly useful when the chain has no homologous counterpart in the database of sequences and may help in highlighting also its function.

Post-translation Modification and Localisation and Results

In BioSapiens-WP8 (2007) work on post-translation modification and localisation, the objectives were to predict protein features, in particular post-translational modifications, localisation signals, and to use combinations of such features to predict cellular role and molecular function for proteins without sequence similarity to proteins of known function. Datasets were constructed and verified for data-driven prediction algorithms, and were made publicly available. One of the tools developed is described in BioSapiens-WP8 (2007) deliverable De8.8, "A neural network based method for prediction of protein localisation to the nucleolar proteome". The nucleolus is the most prominent substructure of the nucleus. To predict nucleolar protein localisation, different data sources were integrated using a semiautomated neural network scheme which was later used to assign and rank nucleolus proteins to highly connected protein complexes. The procedure has an exploratory part and an evaluation part. The exploratory part consists of protein interaction database mining–building interaction complexes with a compiled list of the human nucleolus proteome as "seed." Subsequently, a "reverse" proteomics step is implemented in order to gain experimental evidence of nucleolar localisation of proteins that were not in the seed list but that are predicted to be nucleolar on the basis of their presence in the *in silico* generated high-confidence protein complexes. As the evaluation part of the procedure, a machine-learning method was constructed to produce a score indicating the likelihood that a given *in silico* generated complex is nucleolus-localised. The full list of nucleolus proteins from the top 15 complexes can be found at NUCLEOLUS (2007), and for details, see Hinsby et al. (2006) A picture of the human nucleolus is shown in Fig. 2.4.

In further development of this work, Lage et al. (2007) combined protein–protein interaction data and text mining for disease gene finding in novel ways. They performed a systematic, large-scale analysis of human protein complexes comprising gene products implicated in many different categories of human disease to create a phenome–interactome network. This was done by integrating quality controlled interactions of human proteins with a validated, computationally derived phenotype similarity score, permitting identification of previously unknown complexes likely to be associated with disease. Novel candidates were proposed as being implicated in disorders such as retinitis pigmentosa, epithelial ovarian cancer, inflammatory bowel disease, amyotrophic lateral sclerosis, Alzheimer disease, type 2 diabetes and coronary heart disease.

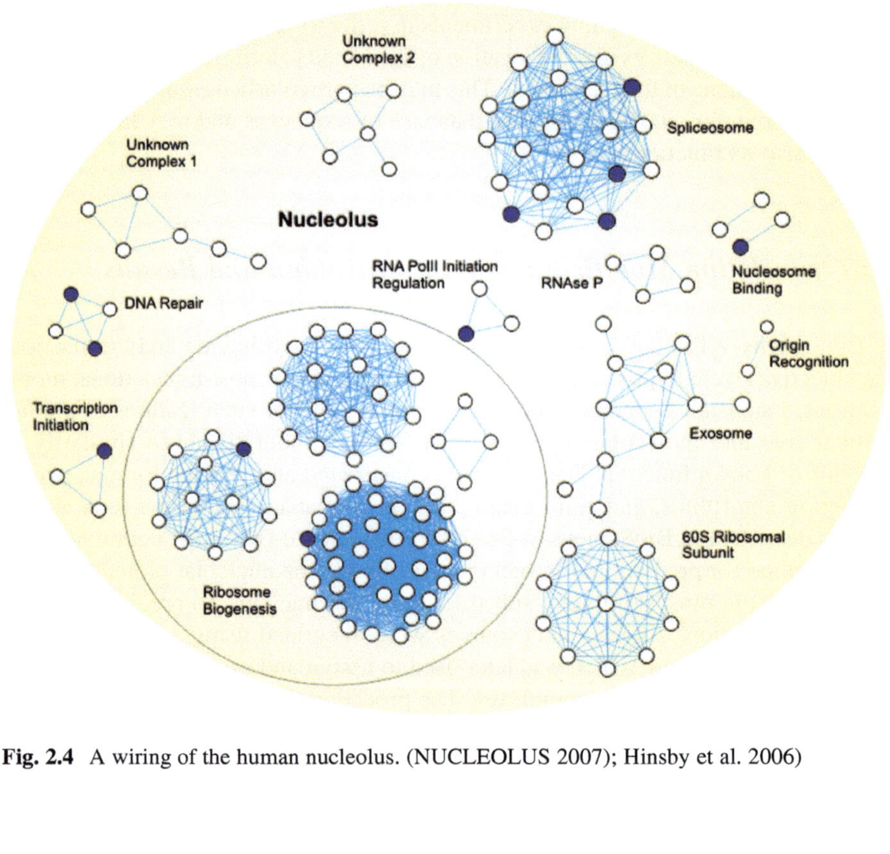

Fig. 2.4 A wiring of the human nucleolus. (NUCLEOLUS 2007); Hinsby et al. 2006)

Future Post-translation Modification and Localisation

Future BioSapiens-WP105 (2007) plans are to predict protein features, in particular post-translational modifications and localisation signals, and to use combinations of such features to predict cellular role and molecular function for proteins without sequence similarity to proteins of known function. It is planned to construct and verify datasets for data-driven prediction algorithms, and to make these publicly available. The work package also includes close link with experiments. A new computational model for transmembrane helices in mammalian proteins will be developed. A Web server that predicts the membrane insertion free energy of peptide segments will be constructed.

Protein Complexes, Networks and Pathways

Protein–Protein Complexes

In BioSapiens-WP10 (2007), the objective is the automatic identification, prediction and analysis of protein interaction partners. The availability of genome sequences and high-throughput biology enables fundamentally different approaches for function prediction. A major resource developed is described in BioSapiens-WP10 (2007) deliverable 10.5, "Integration of experimental data sources into STRING (2007)". The database provides a cross-species protein–protein network of functional interactions. Until recently, however, it only included predicted interactions based on so-called genomic context methods, which greatly limited its usefulness for the analysis of eukaryotic proteomes. BioSapiens deliverable 10.3 describes a unified scoring scheme for experimental and computational protein–protein interactions. STRING (2007) now integrates and scores physical protein–protein interaction evidence from five different databases. The scoring scheme has now been extended to also cover microarray expression data. The latter have been implemented in the form of a Web server called ArrayProspector, which provides microarray-based evidence for STRING (2007). The resulting functional interactions have been benchmarked against the same reference set used for all other evidence types in STRING (2007) to make evidence of different types directly comparable.

Network Prediction

In the area of networks where bioinformatics starts to merge with systems biology, BioSapiens-WP11 (2007) describes moving the field of network prediction and analysis to a status that allows everybody in the community without much effort to exploit the knowledge in the field, with the specific focus of establishing the technology for robust annotation based on network and pathway information for complete genomes. Protocols are described for the combination of the information stored in existing databases on pathways. Procedures are explored for the integration with the pathway information of the network predictions (and the corresponding quality controls). Methods for detailed analysis of the networks, leading to biological discoveries and final functional annotations, are developed.

Metabolic Pathway Net

An important application of this method is shown in BioSapiens-WP11 (2007) deliverable 11.3, "Prediction and annotation of a metabolic net in a model organism". Some bacteria, yeasts, plants, mice, rats and humans utilise the methionine salvage pathway. In this pathway, organic sulphur is salvaged from methylthioribose, which is derived from the methylthioadenosine that is a by-product of the synthesis of spermidine and spermine. This pathway regenerates reduced sulphur and metabolically links it to polyamine biosynthesis, but details of the physiological roles of this pathway remain obscure. The STRING (2007) database employs two different strategies for transferring known and predicted associations between organisms: the first ("COGmode") relies on externally provided orthology assignments and transfers interactions in an all-or-none fashion, whereas the second ("protein mode") uses quantitative sequence similarity searches and often distributes a given interaction fractionally among several proteins of the target organism. With use of protein mode, a network of functional associations was derived for the *Bacillus subtilis* methionine salvage pathway, and for other organisms, based on three types of genomic context evidence. Multiple types of evidence support several of the relations, and they include additional proteins, which are likely to play a role in methionine salvage.

Future Pathway Work

BioSapiens-WP106 (2007) integrates known and predicted protein–protein interactions from a number of databases and prediction methods from the various partners. To be able to do this at a genome-wide scale with the highest possible quality, automation and critical evaluation and scoring of the individual sources are the key principles in this work. While previous work focused on the development of common platforms and capturing of knowledge from existing databases, future activities make use of these resources for predictions and the annotation of more probabilistic features, e.g. the prediction of a missing enzyme or the regulator or transporter with which a pathway is associated. Exploratory work is carried out to capture other cellular processes.

Encyclopaedia of DNA Elements

Functional Elements in the Human Genome

The Encyclopaedia of DNA Elements (ENCODE 2007) was launched in September 2003 by the National Human Genome Research Institute (NHGRI 2007) of the National Institutes of Health (NIH 2007). The goal is to identify all functional elements in the human genome sequence. The pilot phase aims at analysing defined regions of the human genome sequence using existing testing methods and close interactions between computational and experimental scientists. Regions representing approximately 1% (30 Mb) of the human genome have been chosen and were analysed by ENCODE consortium researchers. Fourteen regions were chosen because they were regions of special interest and 30 more regions were chosen randomly from cluster regions that were grouped according to non-exonic conservation and gene density. Whereas this book concentrates mostly on European collaborative research, ENCODE (2007) is an excellent example of a broad collaborative effort based in the USA, with a wide range of international partners (ENCODE-Participants 2007) who are funded, and others with whom major informal collaborations occur at the project level.

International Collaboration

The contribution of BioSapiens (2007) to ENCODE (2007) is vital, especially with the wide range of tools available within the BioSapiens (2007) European Virtual Institute for Annotation. GENCODE (2007) is a BioSapiens-WP20 (2007) and BioSapiens-WP108 (2007) subproject which is associated with ENCODE, which seeks to identify all protein-coding genes in the ENCODE (2007) selected regions.

Functional Identification Methods

Important advances have been made in functional identification by using the full power of the BioSapiens (2007) network for ENCODE (2007), as shown in BioSapiens-WP20 (2007) deliverable De20.1, "BioSapiens–ENCODE collaboration", containing a report describing the status of the work performed. Data are generated by the means shown in Table 2.3, where functional genomic elements are identified. The methods indicated are being used to identify different types of functional elements in the human genome. For each protein-coding gene, the delineation of a complete mRNA () sequence is performed for at least one splice isoform, and often for a number of additional alternative splice forms. Coding sequences for the 44 regions in the study have been ascertained by the Human And Vertebrate Analysis and Annotation (HAVANA 2007) group. In total there are 1,097 coding sequences from the 44 selected regions of the human chromosome. The contributions from the BioSapiens (2007) partners were focused on information from a protein annotation perspective so that, where possible, annotations can be viewed from all groups simultaneously through DAS (2007) servers. Special attention is given to the potential aspect of alternative splicing and the putative effect it has on function by altering domain, structure, localisation and post-translational modification.

Genome Analysis Future

The groups participating in ENCODE (2007) plan to cover 100% of the human genome. BioSapiens (2007) has been participating in the process and the deliverables from this work package will be tailored to the final plan adopted by the ENCODE (2007) partners. A major task is to enable scaling of the protein analysis for full coverage of the human genome, including all the isoforms. The BioSapiens (2007) consortium is particularly interested in the experimental verification of the translation of these genes into proteins.

Future work includes:

- Gene mapping of the 434 gene loci in the set
- Assignment of UniProt sequences, PDB templates, Gene Ontology terms and Pfam domains to the 1,097 sequences in the set
- Comparison of the gene/variants from the randomly and manually selected regions
- Detailed study of a large number of examples from the set in respect of their function and relationship to disease
- Comparison of the supporting evidence for the most interesting splice variants
- Study of the TRANSFAC (2007) sequences from the set

Table 2.3 Indicated methods being used in ENCODE to identify functional elements in the human genome. (From ENCODE-Project-Consortium 2007)

Feature class	Experimental techniques
Transcription	Tiling array, integrated annotation
5′ ends of transcripts	Tag sequencing
Histone modifications	Tiling array
Chromatin structure	Quantitative PCR, tiling array
Sequence-specific features	Tiling array, tag sequencing, promoter assays
Replication	Tiling array
Computational analysis	Computational methods
Comparative sequence analysis	Genomic sequencing, multisequence alignments, computational analysis
Polymorphisms	Resequencing, copy number variation

Major Result: Most DNA Is Transcribed to RNA

A major paper describing the work of the NIH (2007) funded ENCODE (2007) consortium and a number of accompanying papers have been published by the ENCODE-Project-Consortium (2007). The BioSapiens (2007) results are available at GENCODE (2007), see Tress et al. (2007). The findings of ENCODE-Project-Consortium (2007) promise to reshape our understanding of the functioning of the human genome. They challenge the traditional view of our genetic blueprint as a tidy collection of independent genes, pointing instead to a network in which genes, regulatory elements and other types of DNA sequences interact in complex, overlapping ways. In an analysis effort led by the European partners, the ENCODE (2007) consortium's major findings include the discovery that the majority of human DNA is transcribed into RNA and that these transcripts extensively overlap one another. This broad pattern of transcription challenges the long-standing view that the human genome consists of a small set of discrete genes, along with a vast amount of "junk" DNA that is not biologically active. The new data indicate that the genome contains few unused sequences; genes are just one of many types of DNA sequences that have a functional impact. These discoveries are fundamental to the future course of biomedical research.

Chapter 3
Systems Biology

Abstract Systems biology research is presented in this chapter in terms of the biological processes that are investigated. The overall cell cycle is first considered, followed by analysis of the key p53 gene for apoptosis control and an analysis of the key process of spindle formation and related imaging techniques. Signalling and control are at the heart of systems biology analysis, and this is considered in detail for metabolic regulation. The circadian clock provides an excellent example of critically time dependent behaviour. The techniques are then extended to multiple pathway integration, and then even further to cellular systems biology. A major Seventh Framework Programme initiative has led to several new and large projects in several of these areas.

Introduction

Systems Biology Projects

Compared with bioinformatics projects, collaborative research projects in systems biology tend to have a proportionally larger "wet-laboratory" experimental component. This is often because of lack of standardised data or crucial time-dependent data in many areas of research. It is generally necessary to have a strong coupling and constant interaction between model development, computer modelling and data generation from "wet-laboratory" experiments. Even in the largest research institutions, there may not be the right mixture of skills to carry out this necessary interactive model and experimental development. Collaborative research, often involving several laboratories in several countries, becomes necessary for significant advances. Systems biology approaches recognise the importance of wholeness, acknowledging that systems cannot be understood by investigation of their parts in isolation. Today, systems biology brings mathematics, engineering, physics and computer science expertise to the exploration of complex biological systems and their regulation. The current emphasis on systems in biology is the result of recent developments in molecular biology and biochemistry, which have enabled researchers

F.B. Marcus, *Bioinformatics and Systems Biology*,
DOI 10.1007/978-3-540-78353-4,

to collect comprehensive datasets on the performance of systems, and to acquire information about their molecular substrates. Projects discussed include AMPKIN (2007), BaSysBio (2007), BIOSIM (2007), COMBIO (2007), COSBICS (2007), DIAMONDS (2007), EAMNET (2007), ENFIN (2007), EUCLOCK (2007), HepatoSys (2007), QUASI (2007), RiboSys (2007) and SysMO (2007), as well as several FP7 (2007) projects presented at the end of the chapter.

Networks and Dynamics

Human disease phenotypes are controlled not only by individual genes and their products, but also by networks of interactions that exist between those genes and their products, and the system-wide dynamic behaviour that they display. The networks range from metabolic pathways to signalling pathways that regulate hormone action. When perturbed, they alter their output, which, depending on the environmental context, results in either a pathological or a normal phenotype. Study of the dynamics of these networks, using approaches such as metabolic control analysis (for metabolic networks), or stochastic or logical approaches (for gene regulation networks), may provide new insights into the pathogenesis and treatment of complex diseases such as cancer. For example, it is possible to identify groups of genes/proteins which play a critical role in the network. In what follows, the projects and results have been somewhat arbitrarily divided into specialised areas. In fact, there is considerable overlap in these areas, precisely because a systems approach is taken.

Cell Cycle

Cell Cycle Regulation

The DIAMONDS (2007) project focuses on eukaryotic cell cycle regulation, and has developed and implemented computational models of cell cycle control that function as hypothesis-generating engines in a systems biology "wet-laboratory" environment. It has produced a very impressive publication list, and the deliverables give a great deal of detail as to how the work is actually carried out. The work was done in a number of wet laboratories and dry laboratories, on two types of yeast and plant and human cells, to make sure that the approach is validated across widely different organisms. The main target of the project consists of two parts: a cell cycle knowledge base and an integrated platform of data mining, modelling and simulation tools that allows the integrated analysis of that data in a systems biology approach – the development of a basic model, the use of this model to design new experiments, the production and analysis of novel data and the integration of new findings in a more refined model.

Cell Cycle Ontology and Knowledge Warehouse

The major means to reach this target was to harvest and/or produce a large body of cell cycle related biological knowledge. This is functioning as the central resource for the modelling and simulation environment. The knowledge warehouse, designed as the application cell cycle ontology (CCO 2007) constitutes one of the major products of the project, enabling future hypothesis-driven research. CCO contains knowledge of cell cycle related components. The project showcases the fact that a systems biology approach towards analysis of a fundamental biological process has become mature, and hinges on an integrated data analysis pipeline, extended with modelling and simulation tools. Essential elements of such a pipeline are functional genomics data production (transcriptome and proteome), literature mining, comparative analysis of genes and networks, a visualisation, modelling and simulation environment, and a Web service based data integration layer.

Cell Cycle Model Organisms

The core set of model organisms includes:

- *Saccharomyces cerevisiae* (budding yeast)
- *Schizosaccharomyces pombe* (fission yeast)
- *Arabidopsis thaliana* (weed, dicot model plant)
- *Homo sapiens* (human)

Cell Cycle Research Objectives

The objectives are:

- To deliver a knowledge base for future cell cycle control research
- To present a showcase example for a systems biology approach in a variety of eukaryotes
- To design a mathematical model of cell cycle networks on the basis of genome-wide datasets
- To produce a systems biology data integration and modelling environment
- To identify cell cycle targets of major signalling pathways affecting cell division
- To initiate validation of these targets through exploration of perturbations (mutants, small molecules) in wet-laboratory experiments

Systems Biology Data Generation

DIAMONDS (2007) makes use of a number of established technologies developed for genome-wide applications, and combines them such that they assemble a large toolset to perform a comprehensive mining of a broad array of data types for patterns, annotations and other parameters embedded and associated with these. Technology applications include the analysis of the transcriptome, targeted proteomics approaches (including the analysis of the dynamics of protein modification), and integration with prior knowledge from the literature (text mining approach) and annotated databases. The combined data are integrated into rigorous dynamical models that have been progressively refined through subsequent *in silico* simulations and experimental validation. Again, the core foundation for this already existed, and was further refined and most importantly filled with curated data in the course of the project. The components of the regulatory networks identified were systematically questioned for amenability to chemical perturbation (modification, inhibition or blocking) of the cell cycle, to assess the potential for drug design, their involvement with growth characteristics and in general their fundamental role in the regulatory system.

Cell Cycle and Health

Cell cycle regulation is of particular significance for human health because:

- Disturbances of the cell cycle regulatory network lie at the basis of many cancer types.
- The comparative approach illuminates the variation in the intrinsic stability of cell cycle controls in plants and animals, providing insight into proliferative disorders.
- The analysis of the mode of action of cell cycle regulators provides a basis for identification of potential therapeutic targets.
- The use of a cell cycle simulation model helps to predict the value of putative therapeutic components.

Standard Cell Synchronisation

The DIAMONDS (2007) deliverables illustrate how it is vital to combine experimental and modelling work in the same project, and why a large-scale collaborative project is required. To gather uniform data across time and across model organisms, standard cell synchronisation procedures had to be developed. In deliverable D1.1, "Optimised synchronisation protocol for yeasts, Arabidopsis, human cells", it is shown how this synchronisation is achieved. Deliverable D1.2, "Transcript data", provides a detailed description of transcriptome evolution throughout the cell cycle, and shows how these protocols are used in experiments and how data are collected and modelled, at a level of detail that would not necessarily be available in journal publications.

Periodically Regulated Genes

A wide range of bioinformatics tools are available and being developed for data analysis. A good description is given in deliverable D2.4, "Method for cell cycle protein identification beyond primary sequence similarity", where it is described how a neural network based tool HCYCLEP (2007) has been developed for identifying human cell cycle periodically regulated genes on the basis of protein features, such as isoelectric point and subcellular localisation, and which uses protein sequences as input.

Functional Modules in the Cell Cycle

Data are combined and modelled in sophisticated ways. In DIAMONDS (2007) deliverable D3.4, "Methods for the extraction and analysis of functional modules, pathways", the notions of functional modules are used in various contexts and different meanings in biology. They focus on four aspects:

1. Biological regulatory networks usually encompass different subnetworks assignable to specific function. In the case of the cell cycle, one can distinguish subnetworks involved in the control of the entry, or of the exit of the cell cycle, as well as further subnetworks associated with various checkpoints. Encompassing specific regulatory components and their cross-regulation are cross-regulatory modules. Their identification and analysis have been addressed using a graph-theoretical representation and graph analysis algorithms.
2. Modules can also be defined in terms of physical association between regulatory components. In the case of the cell cycle, various complexes are formed at phases of the cell cycle. Such dynamic modules have been studied by systematically analysing a combination of transcriptome and proteome data to map complex formation and activities along the cell cycle.
3. Temporal interplay between regulatory products can also be assessed on the basis of model simulation. A graph-based formalism is used to represent the cell cycle dynamics and to apply graph analysis algorithms to delineate crucial features, such as dynamical cycles.
4. The term "module" is also often used to refer to contiguous DNA regions (promoters, enhancers) involved in the *cis*-regulation of a given gene and driving its expression at a specific location (cells, tissue, etc.) or at a specific time. A series of tools have been implemented to delineate these *cis*-regulatory modules.

Data Visualisation and Analysis

For visualisation of all the data and models, various tools have been developed and are described in DIAMONDS (2007) deliverable D4.1, "Graph edition and visualisation software for qualitative model building". The core of the data analysis system is composed of Expression-Profiler (2007), the CCO (2007) knowledge base and TAVERNA (2007) enabled workflow management with various query and visualisation options.

P53

The p53–Mdm2 Regulatory Network

The gene p53 has perhaps the largest number of published papers devoted to it of any gene, nearly 44,000, because of its key role in apoptosis (cell death) and because it is inactivated by mutations in many cancers. By combining experimental, simulation and bioinformatics approaches, COMBIO (2007) aims to increase understanding of two biologically important systems: the first is the p53–Mdm2 regulatory network, in which the oncoprotein Mdm2 controls the activity of the tumour suppressor "gatekeeper" protein p53, via a negative-feedback loop, and the second is the self-organisation process whereby chromatin controls microtubule nucleation and organisation during spindle formation. These two systems have been selected because they represent two important and different kinds of biological system, one which can be described approximately as a network of free components, and the other in which localisation, self-organisation and gradients play an important role. The general objective is to benchmark the ability of current modelling and simulation methods to generate useful hypotheses for experimentalists, and to provide new insights into complex biological processes. In both systems, the p53–Mdm2 regulatory network, and the dynamics of spindle assembly, different approaches are used to obtain quantitative data, as well as data regarding localisation and the dynamics of the system.

Collaborative Approaches to Data Handling

In close collaboration with experimentalists, databases are developed that are adapted to experimental work and computer modelling. Data are stored in such a way that they are accessible to various simulation packages, and are displayed in such a way that non-experts are able to make sense of them. This aspect requires significant technological innovation. COMBIO (2007) has led to the development of modelling approximations while simultaneously conducting experiments to validate the models' predictions. The different modelling tools have been assessed and a handbook (Di Ventura et al. 2006) has been drawn up, allowing the rapid dissemination of these tools to the broader experimental community. The handbook also indicates which simulations and experimental procedures might be combined, and how to answer important questions about biological function.

Negative Feedback: p53 and Mdm2 Experiments

The experimental work was reported (Geva-Zatorsky et al. 2006) as follows: understanding the dynamics and variability of protein circuitry requires accurate measurements in living cells as well as theoretical models. To the negative-feedback loop between the tumour suppressor p53 and the oncogene Mdm2 were investigated. The dynamics of fluorescently tagged p53 and Mdm2 were measured over several days in individual living cells. By green fluorescent protein (GFP) tagging and careful image analysis, one can measure gradients, oscillations and noise. Isogenic cells in the same environment behaved in highly variable ways following DNA-damaging γ-irradiation: some cells showed undamped oscillations for at least 3 days (more than ten peaks). The amplitude of the oscillations was much more variable than the period. Sister cells continued to oscillate in a correlated way after cell division, but lost correlation after about 11 h on average. Other cells showed low-frequency fluctuations that did not resemble oscillations.

Negative Feedback: p53 and Mdm2 Modelling

The corresponding modelling was reported (Ciliberto et al. 2005) as follows: p53 is activated in response to events compromising the genetic integrity of a cell. Recent data show that p53 activity does not increase steadily with genetic damage but rather fluctuates in an oscillatory fashion. Theoretical studies suggest that oscillations can arise from a combination of positive and negative feedbacks or from a long negative feedback loop alone. Both negative and positive feedbacks are present in the p53–Mdm2 network, but it is not known what roles they play in the oscillatory response to DNA damage. A mathematical model was developed of p53 oscillations based on positive and negative feedbacks in the p53–Mdm2 network. According to the model, the system reacts to DNA damage by moving from a stable steady state into a region of stable limit cycles. Oscillations in the model are born with large amplitude, which guarantees an all-or-none response to damage. As p53 oscillates, damage is repaired and the system moves back to a stable steady state with low p53 activity. The model reproduces experimental data in quantitative detail.

Different families of mathematical models of the system were analysed, including a novel checkpoint mechanism. Modelling points to the possible source of the variability in the experimentally observed oscillations: low-frequency noise in protein production rates, rather than noise in other parameters such as degradation rates. This provides a view of the extensive variability of the behaviour of a protein circuit in living human cells, both from cell to cell and in the same cell over time.

Publications

Many of the results and approaches are summarised in a review article (Di Ventura et al. 2006). Simulations, increasingly paired with experiments, are being successfully and routinely used by computational biologists to understand and predict the quantitative behaviour of complex systems, and to drive new experiments. Nevertheless, many experimentalists still consider simulations an esoteric discipline only for initiates. Suspicion towards simulations should dissipate as the limitations and advantages of their application are better appreciated, opening the door to their permanent adoption in everyday research. The overall publication list of the COMBIO (2007) consortium is impressive, with 24 publications from 2004 to 2006, and several in 2007. There have been major advances in understanding the p53–Mdm2 network, in understanding spindle formation and in developing a very wide range of tools suitable for protein–protein interaction and network analysis and computation. The publications and publicly available deliverables demonstrate the power of these collaborative research efforts to tackle problems in a way different from reviewing the work of thousands of individual researchers whose work in the p53 arena has been published.

Spindle Formation and Imaging

Light Microscopy

Imaging techniques have been evolving rapidly, and are now becoming the basis for important work in image processing and applications to systems biology. In the visible microscopy area, major capabilities were pooled and propagated around Europe in FP5 (2007) by the European Advanced Light Microscopy Network (EAMNET 2007), a network of eight European laboratories and two industrial partners working in the field of light microscopy. Their aim is to assist scientists in exploiting the power of imaging by organising practical teaching courses, creating online teaching modules and offering software packages for microscopy. All partners are also members of the European Light Microscopy Initiative (ELMI 2007). EAMNET-Teaching (2007) has large numbers of excellent images, for example of cell motility, where the cellular cytoskeleton is shown at various stages. This type of technology has been applied in COMBIO (2007) studies of spindle formation, using techniques from CDL (2007), where they have stored the COMBIO Movie database, with an example shown in Fig. 3.1. In vivo recording of spindle assembly

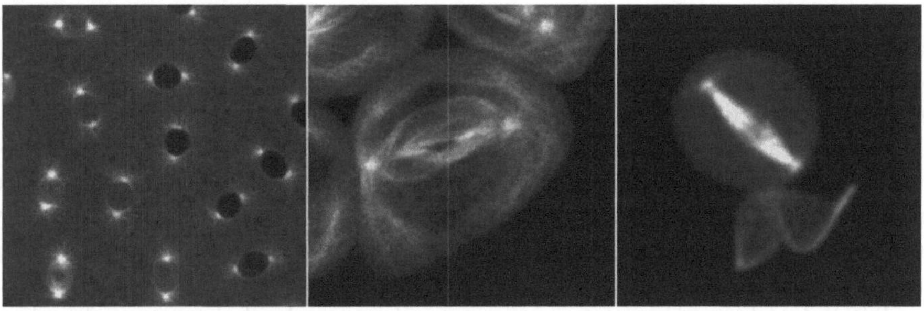

Fig. 3.1 The spindle is a highly dynamic macromolecular structure that self-assembles around the chromosomes. Three different spindle types are shown from *Drosophila*: embryos, spermatocytes and neuroblast, respectively, labelled with green fluorescent protein–α-tubulin constitutively expressed (see COMBIO (2007) and CDL (2007))

is used for the generation of a live imaging repository of spindle assembly movies in different species, and in different cell lineages within a species. This image database is essential for modelling purposes.

Microtubule Formation

Over the last few years, a large number of microtubule-associated proteins have been identified as key players in spindle assembly and function. Collectively they participate in the regulation of microtubule dynamics and organisation, mediating dynamic interactions between the condensed chromosomes and the microtubules. The tight regulation of their activities is particularly important and can happen at different levels. A global regulation is achieved through the activation of the master mitotic kinase cyclin B/cdk1 that changes the cytoplasm from an overall interphase state into a mitotic one. Other finer levels of control modulate locally the activity of microtubule-associated factors. Many enzymes are concentrated directly on the chromatin or other parts of the mitotic spindle, with the consequence of generating local gradients in the activities of proteins involved in microtubule nucleation, stabilisation and organisation. An example is the demonstration that the GTP form of Ran is locally enriched around the chromatin because its exchange factor RCC1 is on the chromosomes, and that Ran controlled several aspects of the microtubule cytoskeleton.

Regulatory Feedback in Microtubule Formation

COMBIO (2007) has examined the mechanism and importance of feedback regulatory loops in microtubule nucleation and organisation around chromatin. The large body of experimental data already available on these processes in different organisms (the molecular components their interactions, binding and kinetic constants, concentrations) was compiled and stored in a database, enabling ready analysis and visualisation. This allows a detailed comparison of processes between organisms, with discovered differences providing new insights or clues for experimental investigations. Measurements were made of gradients, time-lapse image acquisition of mitotic events, GFP localisation, etc. by experimental groups which resulted in the generation of a life imaging repository (Rebollo et al. 2007) of spindle assembly movies in different species and in different cell lineages within a species. Using these data and those compiled from the literature, the theoretical groups developed models and performed simulations to account for the role of chromatin, phosphorylation gradients and component localisation in microtubule nucleation and organisation. Predictions made from the simulations were tested experimentally (Janson et al. 2007).

Signalling and Control

Central Role of Cell Signalling

The research area of cell signalling investigates the transmission of information from receptors to gene activation by means of biochemical reaction pathways that form complex signalling networks and impinge on development and health of organisms. COSBICS (2007) established a novel computational framework in which to investigate dynamic interactions of molecules within cells. Instead of simply mapping proteins in a pathway, COSBICS (2007) is concerned with "dynamic pathway modelling", which establishes mathematical models to quantitatively predict the spatial–temporal response of signalling pathways and subsequent target gene expression. To understand how biochemical networks make decisions, studying the dynamic interactions of proteins is important, rather than just creating static maps of the components involved. Mathematical modelling provides practical, useful tools to design experiments and allows hypotheses testing to generate new biological knowledge. For this, a multidisciplinary approach combining an iterative modelling process with advances in quantitative data generation is essential. This requires close interaction between experimental groups, data analysts and modellers.

Cell Growth, Differentiation and Survival Pathways

The aim was to develop methods that are generic in the sense that they are applicable to signalling networks in general, and as independent of the organism as possible. Towards this end, COSBICS (2007) considered two important systems: the Ras/Raf/MEK/ERK pathway and the JAK/STAT pathway, allowing investigation of the heart of the intracellular communication network that governs cell growth, differentiation and survival. Cancer can be considered a disease of communication at molecular level. Combining mathematical modelling with biology, COSBICS (2007) has improved our understanding of how these two central communication networks are subverted in tumour cells and thus promote the generation of new knowledge in functional genomics.

The Ras/Raf/Mek/ERK Pathway

The Ras/Raf/MEK/ERK pathway shown in Fig. 3.2 is regulated by protein interactions and embedded in signalling networks.

A crucial regulator of this pathway is Raf kinase inhibitor protein (RKIP). COSBICS (2007) has obtained important results that have been published in the following areas:

- Outline of a strategy for power-law modelling of cell signalling systems.
- Implementation and publication of a free power-law toolbox.
- Quantitative datasets of EpoR, JAK2, STAT5, CIS and SHP-1 of sufficient quality to be used in mathematical modelling.
- Quantitative datasets of EpoR, JAK2, STAT5, CIS and SHP-1 in cells with upregulated levels of CIS or SHP-1 of sufficient quality to be used in mathematical modelling.
- An adherent cell line for microscopic studies of the EpoR/JAK2/STAT5 pathway.
- Functional GFP-tagged STAT5, EpoR and ERK1.
- The SILAC liquid chromatography–tandem mass spectrometry experimental identification of new potential binding partners of RKIP.
- Results of the ELISAs together with data from quantitative western blotting, which indicate a low effect of RKIP on the ERK pathway in mammary epithelial cells.
- The downregulation of RKIP, which affects the NF B pathway in breast epithelial cells.
- The successful generation of MCF10A cells with very low level of endogenous RKIP expression by stable introduction of micro RNA against RKIP.
- A B-Raf mutants phosphorylation site mutant was generated in order to test whether this phosphorylation regulates the binding of RKIP to B-Raf.
- A parameter estimation toolbox, allowing models in different formats, including SBML, Fortran and MATLAB.
- Improved optimisation procedures, which have led to a publication in *BMC Bioinformatics*.
- Optimal experimental design toolbox, extended and tested with applications for the JAK2/STAT5 and Ras/RAF/MEK/ERK pathways.
- An RNA-silencing model, which considers a time delay of RISC–messenger RNA (mRNA) complex regeneration.

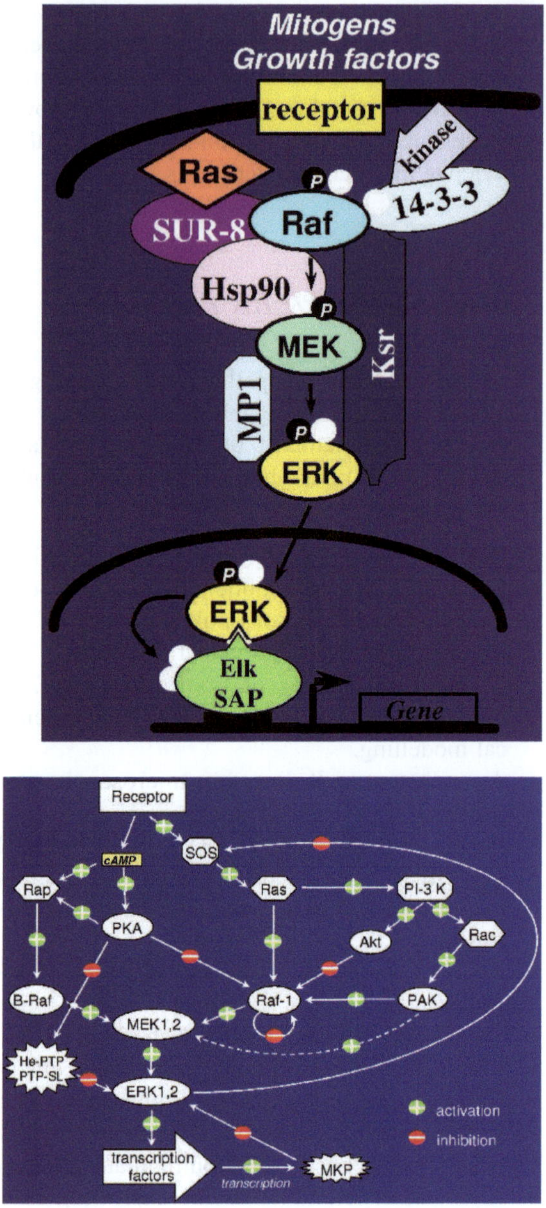

Fig. 3.2 The Ras/Raf/Mek/ERK pathway. See COSBICS (2007)

- Development of a strategy for modelling ERK and STAT crosstalk, by using reaction diffusion distributed models.
- Determination and analysis of a distributed model of ERK and STAT protein interaction.
- Data-based algorithm for identifiability analysis.

- Stochastic integration algorithm to average dynamic functions, e.g. sensitivities, by marginalising non-identifiable parameters.

Data and Modelling Interaction

COSBICS (2007) has entered its most productive phase with a very satisfactory integration of data generation in the laboratory and mathematical modelling. In parallel to the experimental effort, novel theoretical approaches have been developed to identify models from data, to encode models using different mathematical formalisms and to analyse dynamic properties of models. The COSBICS (2007) project is characterised by a wide range of methods, covering the generation of quantitative time course data of two cell signalling systems to the development and application of advanced mathematical and computational techniques.

Quantifying Signal Transduction

QUASI (2007), aims at a better understanding of the systems-level dynamic operation of signalling pathways. Signal transduction pathways are the cellular information routes with which cells monitor their surrounding as well as their own state and adjust to environmental changes or hormonal stimuli. Signalling encompasses the processes with which cells sense changes, generate intracellular signals, transduce the signal and ultimately mount a response. In doing so, signal transduction pathways orchestrate cellular metabolism, establish stress tolerance, control growth, proliferation and development and determine morphogenesis. Consequently, signal transduction pathways are critically involved in disease processes. QUASI (2007) moves from identification of novel components of a system and the detailed molecular and biochemical analyses of individual components to an understanding of the interaction of the components of a system and finally the operation of the system as a whole.

Mitogen-Activated Protein Kinase Pathways

While some standard techniques of postgenomic research, such as microarrays, are suitable tools for experimental studies, additional advanced approaches have to be considered in order to capture time-dependent and spatial effects of the system under study. The budding yeast *Saccharomyces cerevisiae* has been used to study fundamental aspects of cell and molecular biology for more than 40 years. Work on yeast cell cycle and protein targeting has been awarded Nobel prizes in medicine and recently studies on the secretory pathway have received the Lasker prize, illustrating the importance of the model system. Especially in signal transduction through mitogen-activated protein

(MAP) kinase pathways, research on budding yeast, employing the power of genetics and functional genomics, has revealed several important aspects of these pathways, such as the existence of scaffold proteins. The yeast pheromone response pathway and osmosensing high osmolarity glycerol (HOG) pathway are arguably the best understood MAP kinase pathways in general. The amount of available and accessible information is unique and ever-growing. It is, therefore, reasonable and advisable to build on this unique knowledge resource as well as the genetic tractability to advance understanding of signalling in this model system.

AMP-Activated Protein Kinase Signalling Pathway

In AMPKIN (2007), dealing with the systems biology of the AMP-activated protein kinase (AMPK) pathway, experimental and theoretical studies will be integrated to achieve a better understanding of the dynamic operation of the AMPK signalling pathway. This pathway plays a central role in monitoring the cellular energy status and controlling energy production and consumption. Mathematical descriptions will be generated of pathway activation/deactivation in yeast and mammalian cells, as shown in Fig. 3.3. The computational models support drug development in obesity and type-2 diabetes, by employing systems biology in drug target identification and in drug development. AMPK is the sensor of the cellular energy status. The organisation, function and physiological roles of the AMPK pathway are highly conserved from yeast to human AMPK/SNF1 pathways. The detailed molecular mechanisms that control AMPK are still incompletely understood. The central cellular function of AMPK is to switch off ATP-consuming processes and to stimulate ATP production. AMPK also seems to have roles in the control of whole-body energy homeostasis. In yeast AMPK/Snf1 is best known for its role in glucose repression/derepression.

Health Applications of AMPK Modelling

The number of "new molecular entities" submitted to regulatory authorities for registration as therapeutic agents has steadily decreased and most large pharmaceutical companies, as well as the companies in the biotechnology sector, have disappointingly thin drug development pipelines. It appears that an inability to translate genome information into an understanding of biological complexity hampers the development of novel therapies. For this reason, early drug discovery stages will benefit by placing target molecules and their chemical modulators into a meaningful biological context by systems biology approaches. Systems biology approaches can have an impact on development of new disease diagnosis tools and predictions of disease progression, by the holistic understanding of the influence of different genetic and chemical factors on AMPK-dependent energy metabolism.

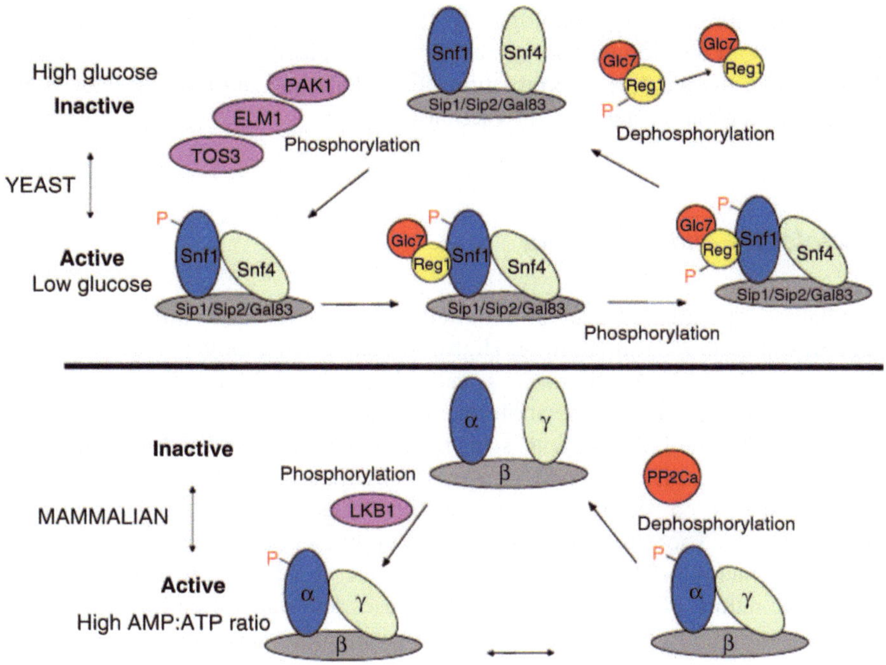

Fig. 3.3 The AMP-activated protein kinase signalling pathway activation and deactivation in yeast and mammalian cells. See AMPKOV (2007)

Metabolic Regulation

Bacillus subtilis as a Model Organism

BaSysBio (2007), a project which started November 2006, uses the model bacterium *Bacillus subtilis* to gain insight into the global structure of the regulatory networks that control bacterial metabolism. BaSysBio aims to understand the regulation of gene transcription in bacteria on a global scale. The highly dynamic gene regulation is mediated by transcription factors, which trigger or repress the expression of their target genes. Transcription control is embedded into a hierarchical flow of information from genes to phenotype, in which many regulatory steps occur.

Regulation of Transcription in Bacteria

Quantitative data need to be generated about the network components at all the levels of the information flow, in order to understand, at the system level, the global regulation of gene transcription in bacteria. This can be achieved by developing and adapting high-throughput technologies to facilitate quantitative measurements, in conjunction with developing and validating computational systems biology methods, to enable quantitative interpretation of the data and unravel the underlying principles of regulatory network interactions.

Technologies and Modelling

Quantitative experimental data are required for each step of the information flow to enable computational modelling. High-throughput technologies (such as living cell arrays, tiling DNA microarrays, multidimensional liquid chromatography proteomics and quantitative metabolomics) are required, in conjunction with new computational modelling concepts, so as to facilitate the understanding of biological complexity. In addition, models need to simulate the cellular transcriptional responses to environmental changes, and their impact on metabolism and proteome dynamics. The iterative process of model predictions and model-driven targeted experiments will refine models, generate novel hypotheses about the mechanistic nature of dynamic cellular responses and unravel emerging systems properties, ultimately providing an efficient roadmap to assist in tackling novel, pathogenic organisms. This system-based strategy can lead to the understanding of how transcriptional regulation and metabolism are quantitatively integrated at a global level, and to understand cellular transcriptional responses in conditions mimicking pathogenesis. The modelling–experimental strategy developed in the highly tractable *B. subtilis* model, when validated, can lead to an understanding of regulatory networks controlling pathogenesis in disease-causing bacteria. At a technological level, BaSysBio (2007) aims to develop and adapt high-throughput technologies for the quantitative determination of the cellular transcriptional responses, to standardise genetic and environmental perturbations, as a function of time, and to develop new concepts in computational modelling and simulation of regulatory networks.

Systems Biology Approach to Transcriptional Modelling

The detailed research strategy involves the following activities:

- Using a novel multipurpose DNA tiling microarray to identify, in a systematic and unbiased way, all the RNA transcripts (mRNAs and small RNAs) produced in the *Bacillus subtilis* cells, and to facilitate a comprehensive inventory of the *cis*-acting regulatory sequences bound by transcription factors
- Bridging technological gaps by developing living cell arrays which allow the genome-wide determination of promoter activities as a function of time during the cell responses

- Exploiting the latest developments in mass spectrometry and non-gel-based protein separation techniques, to quantify proteins and determine their modifications in response to perturbations
- Developing methods for quantitative high-throughput metabolomics, using complementary mass spectrometry based approaches, e.g. gas chromatography–time-of-flight analysis, liquid chromatography (capillary electrophoresis)–electrospray ionisation–time-of-flight analysis and liquid chromatography–tandem mass spectrometry, to analyse the vast chemical diversity of intracellular metabolites in response to perturbations
- Extending the use of parallel ^{13}C-flux analyses to novel substrates
- Developing chromosome engineering tools, based on the recombination systems of prophages of Gram-positive bacteria, to facilitate high-throughput tagging of genes in *Bacillus subtilis* and related pathogens
- Developing new concepts and methods to improve modelling and simulation of regulatory networks. This includes standardised and unequivocal representation of the networks' basic components and interactions to be modelled; hybrid mathematical models combining constraint-based approaches and detailed dynamic modelling.

RNA Metabolism

RiboSys (2007) will use systems biology approaches to model pre-mRNA and pre-ribosomal RNA (rRNA) metabolism in *Saccharomyces cerevisiae* and so aid understanding of these complex cellular pathways. The project plans to:

- Quantify mRNA and rRNA precursors and directly determine rates for their transcription and processing or degradation through the various post-transcriptional pathways
- Produce two comparable mathematical representations of the processing and degradation of pre-mRNAs and pre-rRNAs, and populate the parameters using quantitative experimental data
- Manipulate the model parameters and make predictions about the behaviour of the systems
- Test the predictions experimentally, using yeast mutants that block specific steps
- Use tiling microarrays to investigate antisense and intergenic transcripts, analyse correlations in their expression patterns, and the effects of mutations in transcription, splicing and RNA turnover factors on their transcription and stability
- Use refined imaging techniques to visualise individual transcripts to determine whether the population data reflect the situation in individual cells
- Develop a notation system that permits RNA molecules to be described in a universal format, comparable between different species and organisms, and that is also compatible with a mathematical description

Applications of RNA Metabolic Analysis

Quantitative analyses illustrate the relationships between different steps, activities and factors in the pathway more clearly than has been achieved by qualitative analyses and intuitive interpretations, leading to fresh insights into, for example, the key steps at which regulation would most likely be exerted, leading to testable hypotheses which can be addressed experimentally. Comparison of the pre-mRNA and pre-rRNA models enriches understanding of each pathway, providing further insights into equivalent pathways in human cells, which are less amenable to direct experimentation, thereby enhancing understanding of human genetic disorders.

Circadian Clock

Nature of the Circadian Clock

Behaviour, physiological processes and their biochemistry are temporally structured, and therefore generate daily oscillations. These cycles are not driven simply by external changes (such as the changes of light/dark or warm/cold), but are controlled by an endogenous clock that exists in the most diverse organisms, from cyanobacteria to humans. In real life, this circadian clock is synchronised with the outside world by rhythmic environmental signals through a process called entrainment. Circadian rhythms exist at all levels of biology. They are present, for example, in rest, arousal or vigilance activities; in temperature, urinary output, blood pressure or heart rate; in enzyme activity, hormone concentrations or gene expression. Previous experiments have shown that circadian rhythms continue even in the absence of environmental time cues. A critical feature of the clock is its synchronisation with the external day. This so-called entrainment is the key to understanding the circadian clock and its control mechanisms.

Entrainment

EUCLOCK (2007) aims to investigate the circadian clock in different organisms from cells to humans, and to understand how circadian clocks synchronise with their cyclic environment in the context of entrainment. A major objective of the project is to enable large-scale, non-invasive studies that can prove or disprove the efficacy of medical treatment of pathological conditions, ranging from heart diseases to cancer, using 24-h monitoring of the impact of these treatments, by comparing genetic model organisms and humans, and by identifying new genetic components that control the circadian clock and its entrainment. An integrated systems biology approach is essential to the understanding of the dynamics of the phenomena involved. Protocols, devices and algorithms will be developed, enabling, for the first time, large-scale, non-invasive research on human entrainment in the field.

BIOSIM (2007) also has extensive work in this area; see Chap. 7.

Multiple Pathway Integration

Resources for Systems Biology

ENFIN (2007) will create the next generation of informatics resources for systems biology with a strong focus on the understanding of cell division. The analysis methods are integrated as part of the ENFIN analysis layer and available from the ENFIN core Web service.

Determining Protein Function from Sequence

A different avenue to the prediction of function is the integration of basic sequence features, (e.g. phosphorylation, glycosylation, predicted structure, localisation signals) in computational systems able to predict general classes of protein function, such as DNA binding, transport protein and others.

Functional Sites via Structural Recognition

ENFIN (2007) will develop a library of three-dimensional templates of sites, from "classical" examples such as P-loops for nucleotide binding through to more challenging cases such as exposed proline loops for SH3 recognition. The project will thread multiple alignments focused on proteins of interest onto these structural templates to determine whether a site of interest has been found. De novo site prediction efforts will concentrate on the development of a new generation of glycosylation and phosphorylation predictors, and their application in the context of two experimental systems associated with the network. In both cases, advanced machine classification systems will be used to predict N-, O- and P-linked glycosylation sites and also phosphorylation sites. These predictions will then be compared with experimental data from mass spectrometry analysis in either *Trypanosoma brucei* (glycosylation) or *Homo sapiens* (phosphorylation) proteins.

The experiments will provide data back to the computational investigators and in particular to investigate false-negative predictions (i.e. where there was an experimental modification with no prediction) as this is the most informative class of discrepancy between the prediction and the result. The phosphorylation data will be generated by newly developed procedures for isolating functionally relevant multiprotein complexes from dividing human cells with phosphopeptide isolation techniques and nano liquid chromatography–tandem mass spectrometry (Beausoleil et al. 2004). The glycoprotein analysis will be from a *T. brucei* source via an established protocol of total glycoprotein solubilisation in sodium dodecyl sulphate/urea followed by dilution, lectin affinity chromatography, PNGaseF digestion, tryptic digestion and nano liquid chromatography–electrospray ionization tandem mass spectrometry and nano liquid chromatography–matrix assisted laser desorption/ionisation–tandem time-of-flight analysis (Atrih et al. 2005). This will establish an extensive dataset of in vivo modification sites, which is stored in the ENFIN core using the PRIDE component. The training and prediction machinery will be incorporated into the ENFIN analysis suite. A similar technique will also be applicable to a number of other post-translation modifications (e.g. myristylation). The close collaboration between the experimentally placed and computational groups will form the next set of modification types to be targeted using similar methods.

Exploitation of Features

Having made predictions on a per-sequence basis, work will concentrate on using the set of features on proteins to help understand and place proteins and their modifications in context of an overall process. Preliminary work (Lichtenberg et al. 2003) has shown a surprisingly high signal-to-noise ratio of this "feature space" of protein sequences in the yeast cell cycle. The work will be translated to the mammalian cell cycle system, and will include other feature predictions such as transferase-specific glycosylation and kinase-specific phosphorylation and will also apply the general approach to other pathways, such as endocytosis. The resulting predictions will suggest possible roles of proteins in these pathways and possible critical modification sites. These two types of predictions will be tested by both small interfering RNA and biochemical approaches in these systems. In the cases where there are available antibodies for the predicted proteins, their localisation will be tested before and after endocytosis. For predictions of the association of specific phosphopeptide sites with specific phosphatases, substrate specificity analysis techniques will be used with synthetic peptides (Wälchli et al. 2004). Finally, low-throughput RNA interference techniques will test subsets of predictions using cellular imaging techniques to read out cell cycle or endocytosis related phenotypes. Again, the prediction tools will be made part of the ENFIN analysis suite and thus will be easily applied to other pathway information.

Regulation, Transcription and Signalling

To increase understanding of the interplay between *cis*-regulatory regions of genes, transcriptional regulation and signalling pathways, the approach is to combine *cis*-regulatory network analysis with protein–protein interactions, partial knowledge of signalling networks and comparative genomics to provide partial network reconstructions of pathways of interest. The first aim is to develop and utilise *cis*-regulatory network analysis approaches to identify transcriptional modules. The predictions will be tested in specific signalling pathways, namely the TGF-pathway. The second aim will use protein–protein interaction data and partial network data to predict new genes of interest in a particular pathway using all available data, including density of protein–protein interactions and putative co-regulation inferred from upstream regulatory regions. The methods will use the premature senescence LKB1, and apoptotic pathways as a test bed. The final aim will be to use comparative genomics to compare and reconstruct pathways from diverse eukaryotes. This will be applied to the cell cycle and mitotic pathways, looking principally at yeast and human models. In all cases the methods will be designed to work generically with the ENFIN core system.

Use of Gene Expression Data

The growing availability of gene expression and DNA sequence data creates an opportunity to reveal the molecular mechanisms that regulate the expression of genes on a genome-wide scale. Given the upstream regions of all the genes, and measurements of their expression under defined conditions, it is possible to "reverse engineer" the underlying regulatory mechanisms and identify "transcriptional modules", which are sets of genes that are co-regulated under these conditions through a common motif or combinations of motifs. The TGF-β pathway represents an interesting challenge in regard of transcriptional regulation, as it provides a very complex pathway in terms of its diversity of physiological effects and its dependence on cellular context and alternative signalling context. The intracellular signalling machinery seems to consist of mainly two very closely related Smad cascades, the TGF-β and the BMP Smad cascades. The goal is to identify the set of transcriptional modules that mediate TGF-β signalling specificity, and how these modules change under different conditions.

Systems Modelling

In silico models of pathways of interest make it possible to virtually study the requirement of each network component and to identify the key control elements. Models can either guide experimentalists to chose the molecules to be targeted in priority in the pathway, or further allow testing of very large networks via *in silico* conditions which are not available at the laboratory bench. By integrating the access to both Reactome (2007), a curated database of pathways, and its semiautomatic export-to-model function, and to BioModels (2007), a database of models, ENFIN (2007) provides biologists with a portal for modelling their pathways of interest.

International Collaborations on Systems Biology Tool Development

ENFIN has collaborated since 2006 with the team of Reactome (2007) to optimise the export of curated pathways in a format suitable for further modelling by different methods, such as kinetic modelling by ordinary differential equations or Boolean modelling. ENFIN (2007) has teamed up with the DREAM project (DREAM/ENFIN, 2007) to organise joint workshops to debate the methods to assess computational approaches in the different domains of systems biology.

Cellular Systems Biology

Intergovernmental Collaboration on Bacterial Systems Biology

In addition to BaSysBio (2007), the SysMO (2007) goal is to establish a systemic understanding of key microorganisms with the aid of data-based mathematical modelling. Owing to its ambitious goals SysMO was established as an intergovernmental European transnational funding initiative. European funding agencies were invited to participate and the framework for funding of SysMO was agreed in 2005, followed by publication of national calls in late 2005, and projects starting in 2007. The projects cover different fields of interests:

- BaCell-SysMO – the transition from growing to non-growing *Bacillus subtilis* cells – a systems biology approach
- COSMIC – systems biology of *Clostridium acetobutylicum* – a possible answer to dwindling crude oil reserves
- 3SUMO – systems understanding of microbial oxygen responses
- Ion and solute homeostasis in enteric bacteria – an integrated view generated from the interface of modelling and biological experimentation
- Comparative systems biology – lactic acid bacteria
- PSYSMO – systems analysis of biotechnology induced stresses: towards a quantum increase in process performance in the cell factory *Pseudomonas putida*
- Systems biology of a genetically engineered *Pseudomonas* fluorescence with inducible exo-polysaccharide production: analysis of the dynamics and robustness of metabolic network
- MOSES – microorganism systems biology: energy and *Saccharomyces cerevisiae*

- TRANSLUCENT – gene interaction networks and models of cation homeostasis in *Saccharomyces cerevisiae*
- Global metabolic switching in *Streptomyces coelicolor*
- Silicon cell model for the central carbohydrate metabolism of the archaeon *Sulfolobus solfataricus* under temperature variation

National Programmes on Cellular Systems Biology

From January 2004 on, the German BMBF has funded an innovative and interdisciplinary research initiative in systems biology to complement activities in the German Federal Government's programme "Biotechnology – using and shaping its opportunities". The funding priority HepatoSys (2007) focuses on a quantitative understanding of complex and dynamic cellular processes. The aim is both to arrive at a holistic understanding of these life processes and to be able to present and make these processes accessible *in silico*, i.e. through software, on the computer. HepatoSys consists of four networks and two platforms. The networks focus on modelling of:

1. Detoxification
2. Endocytosis
3. Iron regulation
4. Regeneration

The networks closely interact with each other and with the two platforms:

1. "Cell biology" is responsible for in vitro systems with primary hepatocytes characterisation and manipulations, such as small interfering RNA knockdown.
2. "Modelling" provides the networks with software tools and modelling techniques for signalling pathways and the central metabolism of hepatocytes.

Another important reference programme is UK-Sysbio (2007). An example of this programme is the research at the Oxford Centre for Integrative Systems Biology (OCISB 2007), which is generating quantitatively predictive models of tractable, well-defined, biological problems. The aim is to understand, predict and control physiological behaviour by integrating knowledge of interactions at molecular, cellular and population levels.

Local and Worldwide Scientific Collaboration

National Programmes

The website of EurSysBio (2007) provides links to some of the systems biology websites in Europe and the world. Institutions such as the Centre for Biological Sequence Analysis – Danish Technical University (CBS-DTU 2007) maintain websites with very extensive links (CBS-DTU-Biolinks 2007) to bioinformatics capabilities, as follows:

1. Databases over databases
2. Databases
   - Major public sequence databases
   - Specialised databases
3. Sequence similarity searches
4. Alignment
   - Pairwise sequence and structure alignment
   - Multiple alignment and phylogeny
5. Selected prediction servers
   - Prediction of protein structure from sequence
   - Gene finding and intron splice site prediction
   - Other prediction servers
6. Molecular biology software links
7. Ph.D. courses over the Internet
8. HMM/NN (Hidden Markov model/neural network) simulator
9. Bioinformatics-related meetings and conferences

International Survey of Systems Biology

Similarly, a recent survey (Cassman et al. 2007) provides excellent descriptions of programmes at individual laboratories, and on collaborative research programmes within countries. One of the most ambitious is the HepatoSys (2007) German Network Systems Biology Hepatocyte programme, described in detail by Cassman et al. (2007) and above. Another important reference programme is the UK-Sysbio (2007) UK's Integrative Systems Biology: BSSRC + EPSRC programme.

European Intergovernmental Programmes

SysMO (2007), is a European transnational funding and research initiative on systems biology of microorganisms, as discussed above.

USA Glue Grants

Cassman et al. (2007) indicate that formally funded large-scale collaborative programmes in the area of systems biology in the USA are relatively rare. They refer to the example of Glue-Grants (2007). The purpose of this initiative is to make resources available for currently funded scientists to form research teams to tackle complex problems that are of central importance to biomedical science and to the mission of the National Institute of General Medical Sciences (NIGMS 2007), but that are beyond the means of any one research group. The NIGMS (2007) supports basic biomedical research that increases understanding of life processes and lays the foundation for advances in disease diagnosis, treatment and prevention. The Institute's programmes encompass the areas of cell biology, biophysics, genetics, developmental biology, pharmacology, physiology, biological chemistry, bioinformatics, computational biology, and minority biomedical research and training. A high level of resources may be requested to allow participating investigators to form a consortium to address the research problem in a comprehensive and highly integrated fashion. There are five of these projects in total:

1. LIPID MAPS Consortium
2. Consortium for Functional Glycomics
3. Inflammation and the Host Response to Injury
4. Cell Migration Consortium
5. Alliance for Cell Signalling (AFCS 2007)

US Integrative Cancer Biology Programme

Another collaborative effort is represented by the Integrative Cancer Biology Programme (ICBP 2007) of the US National Cancer Institute (NCI 2007). In addition to funding a number of interdisciplinary centres, the ICBP (2007) centres interact and collaborate with other NCI programmes and external groups. NCI's Cancer Biomedical Information Grid (CABIG 2007) programme coordinates all the bioinformatics software needed by the ICBP, as part of CABIG's ongoing effort to simplify and integrate the sharing and usage of data by providing access to NCI's cancer research communities.

Informal but Structured International Collaboration

Informal but structured international collaboration is common, where protocols for cooperation and data exchange and analysis are structured, but where each participating organisation has separate rather than collective funding. This is most in evidence in the large international databases, such as genome sequence and protein structure. The EMBL-Bank (2007) Nucleotide Sequence Database (also known as EMBL-Bank) constitutes Europe's primary nucleotide sequence resource. The main sources for DNA and RNA sequences are direct submissions from individual researchers, genome sequencing projects and patent applications. The database is produced in an international collaboration with GenBank (USA) and the DNA Database of Japan (DDBJ). Each of the three groups collects a portion of the total sequence data reported worldwide, and all new and updated database entries are exchanged between the groups on a daily basis. The MSD (2007) – the European project for the collection, management and distribution of data about macromolecular structures – is derived in part from the Protein Data Bank (PDB). The SBML (2007) organisation provides a standard format for systems biology programming and access to the websites of over 100 participating projects. Although funding is separate for each country, steering committees meet from the collaborating countries to set standards and protocols for data exchange. These are very successful, highly focused on one type of data. Even though European Commission funded collaborative research projects have a detailed contract for the participating members, most projects are highly open to outside collaboration with individuals, laboratories and countries, and this occurs very often in practice.

A Major FP7 Initiative in Systems Biology

Projects in Systems Biology Research

The systems biology report (Jehensen and Marcus 2005) provided important input to the European Commission consultation process, which led to a series of related topics in the first call for proposals FP7-CALL-HEALTH-2007-A (2007). Several proposals were received and evaluated, and the following were chosen for contract negotiation within the available budget. On the basis of on past experience, it is very probable although not certain that these proposals will become operating projects, at which time more details will be available in the FP7 (2007) projects catalogue and from the project websites, accessible by searching the Internet for their acronym. The following is The subject (in bold) and the topic published by the Commission (in italics), followed by The project acronym (in bold) and the project abstracts, which are published on the FP7 (2007) projects website.

Unicellular systems: *A system approach to eukaryotic unicellular organism biology.*
UNICELLSYS: Eukaryotic unicellular organism biology – systems biology of the control of cell growth and proliferation. – The overall objective of UNICELLSYS is a quantitative understanding of fundamental characteristics of eukaryotic unicellular organism biology: how cell growth and proliferation are controlled and coordinated by extracellular and intrinsic stimuli. Achieving an understanding of the principles with which bio-molecular systems function requires integrating quantitative experimentation with simulations of dynamic mathematical models. UNICELLSYS brings together a consortium of leading European experimental and computational systems biologists that will study cell growth and proliferation at the levels of cell population, single cell, cellular network, large-scale dynamic systems and functional module. Building computational reconstructions and dynamic models will involve different precise quantitative measurements as well as complementary approaches of mathematical modelling. A major challenge will be the generation of comprehensive dynamic models of the entire control system of cell growth and proliferation, which

will require integration of smaller sub-models and reduction of complexity. Implementation of the models will allow observing responses to altered growth conditions zooming in seamlessly from populations consisting of cells of different replicative age and cell cycle stage via genome-wide molecular networks, large dynamic systems to detailed functional modules. Employing computational simulations combined with experimentation will allow discovering new and emerging principles of bio-molecular organisation and analysing the control mechanisms of cell growth and proliferation. The project will deliver new knowledge on fundamental eukaryotic biology as well as tools for quantitative experimentation and modelling.

Immunology: *Modelling of T-cell activation.*
SYBILLA: Systems Biology of T-cell Activation in Health and Disease. T-cell activation, whether induced by pathogens or auto-antigens, is a complex process relying on multiple layers of tightly controlled intracellular signalling modules that form an intricate network. Defects in this network can cause severe and chronic disorders such as autoimmune diseases. Although 5% of the population suffer from these diseases, only a few therapeutic treatments are available. To a large extent this is attributed to the lack of systems-level insights, which would provide concepts of how to modulate T-cell activation. Through a multidisciplinary effort it aims to understand at the systems' level, how T-cells discriminate foreign from auto-antigens. Towards this goal, a transgenic mouse system will be used as a tractable physiological model. Data will be validated in human T-cells and a humanised mouse model for multiple sclerosis. SYBILLA will develop technological and mathematical tools to generate and integrate high-density quantitative data describing T-cell activation. Proteomics, transcriptomics, metabolomics, imaging and multiplexed biochemical techniques will be applied to obtain holistic maps of T-cell signalling networks and to achieve a quantitative understanding of the network and its regulation in response to different inputs. Building upon their existing network model, constant iterations will be used to develop more robust dynamic models to describe the network's response to perturbations. This will culminate in the generation of a Virtual T-Cell, allowing computer simulation to refine the predictability of physiological and pathophysiological reactions. SYBILLA's impact on EU biopharmaceutical competitiveness will be enormous through identification of new pharmacologic targets, optimised prediction of immunomodulatory drug efficacy, discovery of new concerted biomarkers and improvement of personalised medication for treating autoimmune diseases.

Stem cells: *Fundamental approaches to stem cell differentiation.*
EuroSyStem: European Consortium for Systematic Stem Cell Biology. See the end of Chap. 4 for a full description.

Apoptosis: *Developing an integrated in vitro, in vivo and systems biology modelling approach to understanding apoptosis in the context of health and disease.*
APO-SYS: Apoptosis systems biology applied to cancer and AIDS – An integrated approach of experimental biology, data mining, mathematical modelling, biostatistics, systems engineering and molecular medicine. See the end of Chap. 8 for a full description.

Supporting tools and data for systems approaches: In addition, topics were provided at the same time that would provide further tools and data in support of these system approaches:

Proteomics: *Temporal and spatial proteomics to study biological processes relevant to human health.*

PROSPECTS, PROteomics SPECification in Time and Space. Proteomics is a major new field in biomedical research, which deals with the large-scale identification and characterization of large groups of proteins, or "proteomes". These can either be the components of a subcellular organelle or compartment, or even the entire protein complement of whole cells and tissues. Proteomics is essential in the functional annotation of the genome and in future attempts to build a quantitative, 'systems-based' description of cell biology. However, current 'first generation' proteomics approaches largely measure protein complexes and proteomes as homogeneous and static entities with little or no quantitative annotation. PROSPECTS (PROteomics SPECification in Time and Space) is a proposal by world leaders in this young discipline to make a major advance, both by developing much more powerful instrumentation and by applying novel proteomics methods that will allow us to annotate quantitatively the human proteome with respect to protein localization and dynamics. Complementary technologies, including mass spectrometry, cyro-electron microscopy and cell imaging will be applied in innovative ways to capture transient protein complexes and the spatial and temporal dimensions of entire proteomes. They will develop these new proteomics technologies in a generic fashion to maximize their utility to the wider biomedical community and they will generate comprehensive data sets that will foster many downstream functional studies. Their approaches will also generate unique insights into the molecular basis of multiple forms of human disease, specifically neurodegeneration and other diseases related to folding stress. The multidimensional data sets generated in PROSPECTS will be integrated using advanced data aggregation and machine learning, made available to the scientific community via annotated online public databases and used as a basis for a systems biological modelling of the human proteome, with spatial and temporal resolution within the cell.

Membrane proteins: *Structure-function analysis of membrane-transporters and channels for the identification of potential drug target sites.*

EDICT, The European Drug Initiative on Channels and Transporters. EDICT allies, for the first time, partners with world-class expertise in both the structural and functional characterisation of membrane transporters and channels. State-of-the-art facilities and personnel for X-ray crystallography, Electron Microscopy and Nuclear Magnetic Resonance and the latest high throughput technology, will provide infrastructure for scientists characterising channel and transporter functions in man and pathogenic microorganisms. Their experts in the analyses of all the databases of these membrane proteins and molecular modelling will work with their industrial partners on specific targets chosen for their potential to improve the health of European citizens, increase the competitiveness of European health-related industries and businesses, and address global health issues. EDICT will increase knowledge of biological processes and mechanisms involved in normal health and in specific disease situations, and transpose this knowledge into clinical applications. By combining computational and experimental analyses, existing detailed molecular models of channel and transporter proteins, and novel structures derived by their partners, will be analysed to identify the critical regions constituting drug targets. These basic discoveries will be translated via *in silico* and experimental strategies with our industrial partners into the design of novel drugs that modify activities of the membrane proteins for the benefit of patients. The range of human proteins covered includes potassium channels, anion and cation transporters, neurotransmitter transporters, cation-transporting ATPases, and mitochondrial transporters. Structures of bacterial homologues to the human proteins are exploited to inform the studies of their human counterparts.

Membrane proteins: *Structure-function analysis of membrane-transporters and channels for the identification of potential drug target sites.*

NeuroCypres, Neurotransmitter Cys-loop receptors: structure, function and disease. Cys-loop receptors (CLRs) form a superfamily of structurally related neurotransmitter-gated ion channels, comprising nicotinic acetylcholine, glycine, GABA-A/C and serotonin (5HT3) receptors, crucial to function of the peripheral and central nervous system. CLRs cover a wide spectrum of functions, ranging from muscle contraction to cognitive functions. CLR (mal)function is linked to various disorders, including muscular dystrophies, neurodegenerative diseases, e.g. Alzheimer's and Parkinson's, and neuropsychiatric diseases, e.g. schizophrenia, epilepsy and addiction. CLRs are potentially important drug targets for treatment of disease. However, novel drug discovery strategies call for in depth understanding of ligand binding sites, the structure-function relationships of these receptors and insight into their actions in the nervous system. NeuroCypres assembles the expertise of leading European laboratories to provide a technology workflow, which enables to embark on this next step in CLR structure and function. A major target of this project is to obtain high-resolution X-ray and NMR structures for CLRs and their complexes with diverse ligands, agonists/antagonists, channel blockers and modulators, which will reveal basic mechanisms of receptor functioning from ligand binding to gating and open new avenues to rational drug design. In addition, the project aims at understanding receptor function in the context of the brain, focusing on receptor biosensors, receptor-protein interactions and transgenic models. This major challenge requires application and development of a multidisciplinary workflow of high-throughput (HT) crystallization and HT-electrophysiology technologies, X-ray analysis, NMR and computational modelling, fragment-based drug design, innovative quantitative methods of interaction-proteomics, sensitive methods for visualization of activity and localization of receptors and studies of in vitro and in vivo function in animal models of disease.

Lipidomics: *High throughput analysis of lipids and lipid-protein interactions.*

Lipidomicnet, Lipid droplets as dynamic organelles of fat deposition and release: Translational research towards human disease. Lipids are central to the regulation and control of cellular processes by acting as basic building units for biomembranes, the platforms for the vast majority of cellular functions. Recent developments in lipid mass spectrometry have set the scene for a completely new way to understand the composition of membranes, cells and tissues in space and time by allowing the precise identification and quantification of alterations of the total lipid profile after specific perturbations. In combination with advanced proteome and transcriptome analysis tools and novel imaging techniques using RNA interference, it is now possible to unravel the complex network between lipids, genes and proteins in an integrated lipidomics approach. This project application of the European Lipidomics Initiative will address lipid droplets (LD) as dynamic organelles with regard to composition, metabolism and regulation. LD are the hallmark of energy overload diseases with a major health care impact in Europe. The project will exploit recent advances in lipidomics to establish high-throughput methods to define drugable targets and novel biomarkers related to LD lipid and protein species, their interaction and regulation during assembly, disassembly and storage. Translational research from mouse to man applied to LD pathology is a cornerstone of this project at the interface between research and development. To maximize the value of the assembled data generated throughout the project, "LipidomicNet" as a detailed special purpose Wiki format data base will be developed and integrated into the existing Lipidomics Expertise Platform (LEP) established through the SSA ELife project. ELife collaborates with the NIH initiative LIPID MAPS and the Japanese pendant Lipidbank and is connected to the Danubian Biobank consortium (SSA DanuBiobank) for clinical lipidomics.

Implications of New Projects

At the moment, these eight topics in the integrated programme proposed for systems biology and support projects might be funded by the European Commission at the level of typically €10 million to €12 million each, spread over 4–5 years. If all the negotiations for potentially funded projects are successful, then the following goals could be achieved:

- New levels of integrated understanding of cellular function would result.
- Important advances would be made in the understanding of the human immune system.
- The basis of stem cell differentiation would be on a much more quantitative basis.
- A full range of computational biologists, wet-laboratory scientists and clinicians would have analysed and integrated data relevant to the key process of apoptosis in cancer and in HIV/AIDS.
- A whole new range of proteomics tools would be made available to generate essential knowledge for systems biology research.
- The study of key membrane proteins would be put on a much more quantitative basis for relevance to disease studies.
- The area of lipidomics, which is key to understanding gene and protein interactions, would be strongly advanced.

In addition to the large projects mentioned above, the broadly defined topic for the second call for proposals (FP7-CALL-HEALTH-2007-B, 2007), with a deadline 18 September 2007, should lead to another dozen or so medium-scale systems biology research projects in the following general area:

Multidisciplinary fundamental genomics and molecular biology approaches to study basic biological processes relevant to health and diseases. Projects should be multidisciplinary and should focus on collecting, analysing and applying quantitative data to enable system biological approaches addressing basic biological processes at all appropriate levels of system complexity.

Chapter 4
Developmental Biology and Ageing

Abstract First and foremost, a systems biology approach to developmental biology involves the extension of signalling to the intercellular domain. An ideal and highly tractable model organism for studying this process is *Arabidopsis*. Plant development in important both for itself and as a study of a complex eukaryotic organism. Research into stem cells and developmental processes in other organisms are discussed, especially in terms of the roles of bioinformatics and systems biology. Aspects of the ultimate stage of development, ageing, are examined from several points of view, including the role of mitochondria and nuclear receptors. Implementation in future programmes in the Seventh Framework Programme is discussed.

Introduction

The Importance of Developmental Biology and Ageing

As discussed by Wolpert et al. (2001), developmental biology is at the core of all biology. Although large amounts of data have been gathered, and a remarkable increase in understanding has been achieved, this is an area where a systems biology approach can bring major advances. As pointed out by Carlson (2004), now is a time when it is increasingly possible to provide mechanistic explanations at a certain level for developmental phenomena that in former times could only be described. Developmental mechanisms are also highly relevant to human health, and in unexpected ways. Major aspects of cancer development involve the hijacking of development processes by the tumour, since there are very similar mechanisms involved for tumour invasiveness and normal embryogenesis and wound healing. Several aspects of the ultimate stage of development, ageing, are examined from several points of view, including the role of mitochondria and nuclear receptors. The projects discussed include AGRON-OMICS (2007), Cells-into-organs (2007), CRESCENDO (2007), EuroStemCell (2007) and MiMage (2007).

F.B. Marcus, *Bioinformatics and Systems Biology*,
DOI 10.1007/978-3-540-78353-4, © Springer-Verlag Berlin Heidelberg 2008

Plant Development

Leaf Biology

AGRON-OMICS (2007) is investigating the growth and development of plants, and in particular of the leaf in *Arabidopsis thaliana* with the aim to improve its productivity, and to throw light on the connectedness of molecular mechanisms in a tractable multicellular model organism system. The methods are highly relevant to a general approach to studying systems biology.

Plant Research

Crops supply food, animal feed, chemicals, pharmaceuticals and renewable sources of materials and energy. Plant growth results in biomass accumulation, which, in turn, is the major determinant of crop yield. Despite its importance and complexity, plant growth is, however, a poorly understood trait. Plants evolved multicellular bodies independently from animals and fungi. This evolutionary step, coupled with photosynthesis, explains why plants rely on mechanisms for growth and development that are unique.

Arabidopsis thaliana as a Model Organism

At the present time, *Arabidopsis thaliana* is one of the best plant species for analysis, with the necessary resources accessible for studying complex traits. The typical growth and development of *Arabidopsis* has been accurately described, providing a solid platform on which to base experimental studies of growth processes. *Arabidopsis* also has unparalleled genomics resources, including a high-quality genome sequence and annotation comprising over 30,000 genes, of which 26,000 code for proteins; tagged mutant alleles for 73% of these genes; a choice of DNA arrays to investigate genome transcription; modification and polymorphisms; comprehensive transcriptome, proteome and metabolome atlases; cloned repertoires for functional proteomics; and RNA interference. Furthermore, haplotype maps of unprecedented density for any eukaryote, including humans, are available for 20 *Arabidopsis* ecotypes, helping association mapping. Finally, the genome sequencing of close relatives (*A. lyrata*, *Capsella rubella*) has been launched and will help to improve the accuracy of comparative genome analyses.

Growth Factors

Growth results from a complex network of processes occurring at different organisational levels (whole organism, organ, cell, molecular module, molecule). Some of the key growth factors involved in these processes have been identified in the past decades via (eco)physiology, cell biology and molecular genetics but many more still have to be found. The major challenges are the elucidation of the interaction networks (e.g. macromolecular complexes, cell-to-cell signalling) that constitute each of the different levels of organisation, and the understanding of how these different levels are linked. For example, growth regulators such as auxin and cytokinin, which coordinates the integration of growth at the cell–organ and organ–whole organism interfaces, have been extensively studied in plants for many years. However, although a few components of their biosynthetic and signalling networks have been uncovered, very little is known about how they impinge on the machinery underlying cell expansion and cell division. The main research goals of the project are:

- To investigate systematically the components controlling growth processes in plant cells (genome sequences, proteins, metabolites)
- To understand how they coordinate their action
- To explain quantitative growth phenotypes at the molecular level

Integrated Approach to Leaf Development

The growth process is studied within a common research framework. High-throughput (HTP) quantitative data are generated, defining growth variables, genetic components of growth, the molecular composition of leaves at successive stages of development, molecular interaction networks and small molecules affecting growth. Mathematical and statistical methods to model and predict leaf processes are being developed and tested in close collaboration with computer scientists, statisticians and experimentalists (PSB-UGENT Software 2007). The suite of analytical tools will be exhaustively tested and modified before being made available as a package of integrated systems biology applications and as Web services. The technology platforms at the core of the research programme have been selected to provide quantitative information at all relevant levels of organisation:

- Growth variables recorded at the level of the whole organism
- Organ and cell
- Profiling of the genome
- Transcriptome
- Proteome
- Metabolome

- Protein–protein
- Protein–DNA interaction networks

Developmental Biology Technologies

Technologies can be classified as follows:

- Well-established methods, but only exceptionally applied at this scale to study a single biological system in an integrative framework, requiring standardisation of existing protocols and datasets; they include microarray transcript profiling, HTP real-time polymerase chain reaction, flow cytometry, large-scale recombinational cloning methods, green fluorescent protein fusion subcellular localisation, yeast two-hybrid, tandem affinity purification, mass spectrometry and chromatin immunoprecipitation.
- More advanced profiling techniques, including large-scale single-nucleotide polymorphism genotyping, systematic enzyme profiling of identical samples, isotope tags for relative and absolute quantitation (ITRAQ) for relative protein quantification, cell flow sorting, Fourier transform infrared microspectroscopy and bimolecular fluorescence complementation.
- Novel HTP techniques requiring extensive development and aimed at taking full advantage of *Arabidopsis* as a model species; they include automated leaf structure analysis at cell-level resolution, *in planta* two-hybrid based on antibiotic selection, *Arabidopsis* cell-based assays and high content screening to study systematically the results of genetic (genome-scale) or chemical (library-scale) perturbations.
- Software tools enabling data integration and biological system modelling.

Results

AGRON-OMICS (2007) will yield four types of results:

1. Novel analytical pipelines will be developed to measure cellular processes across multiple levels, including mass spectrometry, remote macroscopic and microscopic imaging and environmental control. These research efforts require the generation of specific informatics infrastructure, common data standards and analytical tools required to capture, store, distribute and analyse HTP data.
2. Novel well-documented integrated software applications will form the basis for a “plant systems biology toolbox”. These applications will be constructed to allow adaptability and integration with pre-existing software, and will be made freely available to other scientists working in systems biology.
3. Transgenic lines, genetic stocks and constructs created or characterised in the project’s framework will be disseminated via stock centres.

4. The project will generate primary data and biological knowledge including the identification of genes/loci and molecules that control growth, and the construction of models that explain how these components interact and function across pathways and processes. The information relative to leaf growth control networks will be exploited to postulate how best to combine inputs to increase plant biomass production via improved germplasm and the use of growth regulators.

Impact

The project will have a significant impact in several research areas. Firstly, the consortium is pioneering systems biology approaches in order to understand biological complexity in the context of a multicellular organism, and across multiple levels of organisation (cells, tissue and whole organism). The tools, techniques and expertise built up in the course of the project may be used to inform research on the complex mechanisms involved in human disease, which result in alteration of cell growth and development. In particular:

- Current knowledge shows that core molecular processes regulating cell proliferation and cytoplasmic growth are conserved between plants and animal cells.
- Progress in the mechanistic analysis of these molecular pathways in plants may contribute fundamental insight into the biology of human cancers. In addition, in-depth knowledge and modelling of specific molecular pathways may result in the potential to develop translational research projects for biomedical purposes (e.g. production of natural compounds for therapeutic use, and production of vaccines against human diseases in plants).
- Growth processes are difficult to characterise in mammalian species at a scale comparable to that which is the target of the project, or without breaching ethical barriers. In this respect, the ability to systematically genotype, phenotype and profile at the molecular level thousands of individual plants is a unique asset of this project and will be of great value in developing similar system-level research in mammals.
- The project may help to reduce the environmental impact of agriculture. Agricultural practices withdraw about 70% of groundwater resources worldwide. In the long-term, irrigation increases soil salinity and leads to the permanent destruction of otherwise fertile soils. Using plants that have improved water-use efficiency will help contain the amount of water consumed by agriculture and mitigate the impacts of irrigation.
- A major goal in plant science is the development of crops as a source of renewable resources and industrial feedstock. In the coming years, 20% of transport energy will hopefully come from renewable resources. As leaves are the primary harvesters of energy, the integrated knowledge of mechanisms controlling metabolism, growth and environmental responses developed in this project will provide a strong foundation for future work in this area.

Stem Cells and Development

Stem Cell Databases and Bioinformatics Tools

The European Commission supports an extensive stem cell and developmental biology research programme, as summarised by Joliff-Botrel and Perrin (2007) in a Sixth Framework Programme (FP6 2007) project catalogue on stem cells. One of the major projects, and also one producing important bioinformatics tool and resources, is EuroStemCell (2007). This research programme is organised into specific areas of stem cell research, with supporting directed projects. Areas with direct relevance to databases and bioinformatics include:

- The development of a prototype European stem cell database and stem cell registry, which will establish a stem cell database (Stem DB) containing a wide range of information about stem cells – from basic biology to clinical applications
- Stem cell bioinformatics, which will facilitate comparative analysis of the stem cell molecular profiling data, and foster bioinformatics collaborations

The other EuroStemCell research areas include:

- Identification and isolation of stem cells
- Lineage analysis and differentiation potential
- Self-renewal and upscaling (for potential applications)
- Control of differentiation
- Applications in neurological disease
- Applications in muscle repair and neuromuscular disease
- Epidermal repair
- The generation of antibodies for stem cell identification
- A forum for ethics and societal issues related to stem cell research
- Clinical roadmap
- Public engagement and outreach

Stem Cell Genomic Data

The FunGenES (Hescheler et al. 2006) consortium has been formed to map the gene subsets involved in pluripotent, lineage-committed and selected differentiated cell types using gene expression profiling and functional screens in the mouse model organism. Although the complete sequence of a mammalian genome defines the information content of each cell, understanding of the selective usage of this information during the development of specific cell types is limited. The fundamental questions that remain to be answered include which are the gene subsets that define the pluripotential self-renewing state of embryonic stem (ES) cells, partially and terminally differentiated developmental states, and how are transitions between these states regulated (lineage commitment)? FunGenES is creating an atlas of mammalian genomes participating in early and late developmental processes. To fulfil the aim, mouse ES cells were used as an *in vitro* developmental model system that is very close to the human system, as they are pluripotent. They can be differentiated through the three major developmental pathways – ectoderm, mesoderm and endoderm – into many committed cell types and can be genetically engineered with relative ease. Knowledge of genetic pathways in mouse ES cell differentiation and development might be translated to human ES cells and the potential development of stem-cell-based therapies.

The Cells-into-organs (2007) project investigates the functional genomics of development and disease, by elucidating molecular and cellular processes underlying specification and differentiation of mesodermally derived organ systems. Developmental genetics and experimental embryology are integrated with modern cell biology and genome-scale analysis, enabling the identification of genes which function in building a specific organ or in a particular aspect of embryogenesis. A major revelation of developmental biology has been the extent to which molecular strategies are redeployed, even during regeneration. Thus, this information is the basic knowledge required for organ and tissue engineering. It is, however, a task which will far exceed the capacity or expertise of any one research group, and which requires the complementary advantages of different vertebrate and invertebrate systems and the combination of multidisciplinary skills, necessitating collaboration. The extensive publication list on the project website shows the wide range of topics investigated and the data generated.

Mitochondria and Ageing

Evolutionarily Conserved Mechanisms

The overall aim of MiMage (2007) is to elaborate the role of mitochondria in ageing and life-span control of biological systems. Of special interest is the discovery and experimental manipulation of evolutionarily conserved mechanisms shared between invertebrate and mammalian model systems. A range of experimental organisms and cell culture systems are being studied. Specific issues addressed by the project are:

- The effect on ageing of modulating the amount of mitochondrial reactive oxygen species (ROS)
- The role of molecular and cellular pathways involved in maintaining a “healthy” population of mitochondria
- The nature and impact of age-related signalling pathways on mitochondrial functions
- The effect of dietary restriction on mitochondrial activity
- The impact of hitherto-unknown age-related mitochondrial functions

Conserved signal transduction pathways influence ageing in various model systems. The main objective is to establish putative cross-talking between these pathways and mitochondrial activity. The role of Ins/IGF-1 signalling on mitochondrial function in *Caenorhabditis elegans* is investigated. As a prerequisite for this analysis, a reliable method for the isolation of intact mitochondria from *C. elegans* is being developed. Mitochondrial activity in terms of mitochondrial potential and respiratory capacity and ROS production is analysed in selected mutants. This is complemented by a thorough characterisation of metabolic pathways in these mutants.

Databases for Model Organisms and Developmental Biology

Two of the key model organisms in developmental biology are *C. elegans* and *Drosophila melanogaster*. Extensive databases have been developed for both, which are references for related experimental studies. For *C. elegans,* the main databases is Wormbase (2007), which provides a full set of genes and related information. For *D. melanogaster*, the main database is FlyBase (2007), a database of *Drosophila* genes and genomes. In addition to genome information, these databases possess a wealth of detail of developmental biology data.

Mitochondria Signalling

The influence of insulinlike growth factor (IGF) and IGF-binding proteins (IGFBP) signalling on mitochondrial function is also studied in cultured human cells. IGFBP-3 is highly overexpressed in senescent human cells and probably contributes to the senescent phenotype. To determine whether IGFBP-3-induced signals trigger alterations in mitochondrial function, human umbilical vascular endothelial cells and fibroblasts are treated with recombinant IGFBP-3. All factors considered, these experiments should reveal if modulating IGF/IGFBP signalling does induce detectable changes in mitochondrial activity (such as increased ROS production) and if this represents a public or a private mechanism. Finally it is ascertained how the mitochondrial small heat shock protein HSP22, as anticipated, mediates changes in stress resistance that are controlled by IGF signalling in *D. melanogaster*. In parallel with the investigations of the Ins/IGF signalling pathway, the influence of Ras signalling on mitochondrial function is analysed in yeast and human cells. The experiments should reveal whether expression of oncogenic Ras, which is associated with reduction in life span, exerts its influence on life span through modulating mitochondrial function and, in particular, mitochondrial ROS production.

Metabolic Processes

Another important task is the characterisation of cellular response to a breakdown of the energy transduction. This phenomenon is well known as retrograde response in yeast, which is defined as a series of nuclear controlled events that are triggered by a loss of mitochondrial function. This loss leads to changes in gene expression that cause a metabolic reprogramming of the cell. Although several yeast genes (the RTG gene family) that are involved in the retrograde response have been identified, the pathway has not been elucidated completely. At present it is unclear whether the retrograde response is conserved in higher eukaryotes. While responses to energy failure can be detected in model organisms, it is currently not clear whether a single conserved pathway exists. To address this question, orthologues of *Podospora anserina* that are equivalent to the known retrograde response genes of *Saccharomyces cerevisiae* are cloned. In the case of success, a functional analysis is initiated. To determine if a retrograde response is activated in human cells in the case of energy failure, the observation is used that in vitro senescence of human fibroblasts involves a strong downregulation of the ATP to AMP ratio and a strong increase in AMP levels, which is a hallmark of energy failure. Several changes in gene expression of human cells have been characterised. The aim is to establish the regulatory pathway that leads from replicative senescence via elevated AMP levels to G1 arrest. Genes emerging from complementary DNA microarray analysis are systematically compared with yeast genes that are involved in the retrograde response. The ultimate goal is to identify human genes that might have a function similar to that of the RTG gene family.

Nuclear Receptors and Ageing

Nuclear Receptors

CRESCENDO (2007) investigates members of the nuclear hormone receptor superfamily (NRs), which are intracellular transcription factors that directly regulate gene expression, generally in response to lipophilic molecules. Their central position as main transducers of signals mediated by small molecules into genomic regulations makes them the pivotal element in understanding development, physiological processes, ageing, disease and organism–environment interaction. Elucidation of the mode of interaction of NRs with the genome is at the core of modern pharmacology and therapy.

Nuclear Receptor Networks

The complexity of molecular and physiological networks in which NRs play an integral part is being increasingly appreciated. Gaining a useful picture of how these networks operate demands integration of information emerging from reductionist approaches with physiological processes. The project was designed to apply common innovative (post)genomic and bioinformatics technologies in cells, model organisms and humans to advance knowledge (and its eventual application) of signalling dynamics, integration of target gene responses and pathophysiological processes, and to explore ways in which this knowledge might be eventually applied to improvements in human health. NRs are transcription factors, implicated from early development through to senescence and acting as molecular integrators and rheostat controls of complex physiological processes (e.g. growth and metabolism, reproduction, brain function). NRs play crucial roles in many signalling networks, and exert their combinatorial actions through interactions with numerous other signalling cascades. To date, 49 mammalian NR-coding genes are known, each giving rise to one or more variants; the project will direct its attention to ten members (including two orphan receptors) of the NR family which have been implicated in the co-ordinated responses of cells and tissues during development and ageing. The major leaps in our understanding of NR signalling mechanisms over the last decade have led to the formulation of several unifying concepts about how NRs regulate transcriptional control, but have yet to fully illuminate how NRs act as molecular sensors integrating the complex processes that determine developmental and ageing programs.

Databases and Bioinformatics Tools

In CRESCENDO-links (2007), a number of useful databases and bioinformatics tools that are essential to research in this field are reported. One important example of a European resource in this area is NuReBase (2007).

Implementation in the Seventh Framework Programme

Stem Cells

EuroSyStem, the European Consortium for Systematic Stem Cell Biology, brings together European research teams to create a unique and world-leading programme in fundamental stem cell biology. By interconnecting complementary biological and computational expertise, the project will drive the generation of new knowledge on the characteristics of normal and abnormal stem cells. It will pave the way for application of systems methods by measuring and modelling stem cell properties and behaviour. Information will be mined from studies in model organisms, but the primary focus is on the paradigmatic mammalian stem cells – haematopoietic, epithelial, neural and embryonic. The project will compare cellular hierarchy, signalling, epigenetics, dysregulation and plasticity. Niche dependence, asymmetric division, transcriptional circuitry and the decision between self-renewal and commitment are linked in a cross-cutting work package. A multidisciplinary approach combines transgenesis, real-time imaging, multiparameter flow cytometry, transcriptomics, RNA interference, proteomics and single cell methods. Small and medium-sized enterprises will contribute to the development of enhanced-resolution quantitative technologies. A platform work package will provide new computational tools and database resources, enabling implementation of novel analytical and modelling approaches. EuroSyStem will engage with and provide a focal point for the European stem cell research community. The targeted collaborations within the EuroSyStem research project will be augmented by federating European research excellence in different tissues and organisms. The project will organise annual symposia, training workshops, summer schools, networking and research opportunities to promote a flourishing basic stem cell research community. This network will foster interaction and synergy, accelerating progress to a deeper and more comprehensive understanding of stem cell properties. In parallel, EuroSyStem will develop Web resources, educational and outreach materials for scientists and the lay community.

Part II
Resources Supporting Bioinformatics and Systems Biology Research

Chapter 5
Databases, Computational Tools and Services

Abstract Collaborative research and infrastructure projects have generated extensive computational resources that are available to the public over the Internet and grids, in terms of databases, bioinformatics and systems biology tools and services. Many of them are available via major centres and gateways such as the European Bioinformatics Institute; others are available via distributed and often cross-linked resource centres. Tools such as the Distributed Annotation System permit experimentalists to contribute data over a wide range of activities. Database grid integration strongly links these resources together in highly useful and operational ways. These databases and resources are supported by ontologies which contribute fundamentally to their organisation and utility, supported by extensive text mining tools. Finally, a full range of systems biology toolboxes are being developed and made generally available not only to computational biologists, but also to experimentalists as targeted users.

Introduction

Computational Resources

Collaborative research projects have generated extensive computational resources that are available to the public over the Internet, in terms of databases, bioinformatics and systems biology tools and services. Many of them are available via major centres and gateways such as the European Bioinformatics Institute (EBI 2007); others are available via distributed and often cross-linked resource centres. Tools such as the Distributed Annotation System (DAS 2007) permit experimentalists to contribute data over a wide range of activities. Database grid integration strongly links these resources together in highly useful and operations ways. These databases and resources are supported by ontologies which contribute fundamentally to their organisation and utility, supported by extensive text mining tools. Finally, a full range of systems biology toolboxes are being developed and made generally available not only to computational biologists, but also to experimentalists as well.

F.B. Marcus, *Bioinformatics and Systems Biology*,
DOI 10.1007/978-3-540-78353-4,

Major Gateway to Collaborative Research

A Central Resource Gateway

EBI (2007), significantly due to European Commission collaborative research and infrastructure funding, has transformed itself from being initially a repository for major bioinformatics databases into a gateway to the collaborative research networks, results, resources, tools and services that link together major resources in bioinformatics and systems biology all over Europe. It is not the only gateway, there are many others, e.g. the Centre for Biological Sequence Analysis – Danish Technical University (CBS-DTU 2007), and other high-quality institutions. However, the EBI (2007) is worthy of special attention, since it:

- Coordinates three major networks: BioSapiens (2007), EMBRACE (2007) and ENFIN (2007)
- Participates in several other projects such as the infrastructure projects TEMBLOR (2007) and FELICS (2007)
- Functions as the repository for many databases and tools, often with worldwide mirror databases
- Develops the tools for accessing the bioinformatics grid

Moreover, many of the collaborative research tools developed have been incorporated into the mainstream EBI capabilities, greatly enhancing the utility of all the coupled resources.

General Search Tool

The standard world tool for database searching is Entrez (2007). The most recent and visible example of integrated capabilities resulting from collaborative research is the EB-eye (2007) search tool on the EBI (2007) homepage. This search tool was developed as a result of the funding and technologies made available via TEMBLOR (2007) and EMBRACE (2007). By inputting, for example, the gene p53 as in Fig. 2.1 for Ensembl and its DAS data, one obtains the following far more extensive output, providing links to all the bioinformatics capabilities listed:

- 178 Genomes
    - 127 Ensembl Selected eukaryotic genomes
    - 51 Integr8 Completed genomes and proteomes
- 6,408 Nucleotide Sequences
    - 5,922 EMBL-Bank Europe's primary nucleotide sequence resource
    - 486 EMBL-Bank (Coding Sequence) Coding Sequences in EMBL-Bank

- 626 Protein Sequences
  - 24 PRIDE Proteomics Identification Database
  - 602 UniProt KB UniProt Knowledge Base of protein sequences
- 128 Macromolecular Structures
  - 128 MSD/PDB Macromolecular structures database
- 1 Small molecules
  - 0 ChEBI Chemical Entities of Biological Interest
  - 0 Ligands Library of ligands, small molecules and monomers
  - 1 RESID Protein residue modifications database
- 56 Gene Expression
  - 7 ArrayExpress (Repository of Microarray data) ArrayExpress Repository is a MIAME compliant public database for microarray data
  - 1 ArrayExpress (Warehouse of Microarray experiments) ArrayExpress Warehouse is an expert-curated database of gene expression profiles
  - 48 ArrayExpress (Warehouse of gene expression profiles) ArrayExpress Warehouse is an expert-curated database of gene expression profiles
- 190 Molecular Interactions
  - 96 IntAct Experiments Experimental procedures used to characterise molecular interactions
  - 84 IntAct Interactions Descriptions of molecular interactions
  - 10 IntAct Interactors Proteins taking part in molecular interactions
- 18 Reactions & Pathways
  - 0 BioModels Database of Mathematical models of biological interest
  - 18 Reactome Database of core biochemical pathways and reactions
- 53 Protein Families
  - 53 InterPro Database of protein families, domains and functional sites
- 1 Enzymes
  - 1 Intenz Integrated relational Enzyme database
- 42,963 Literature
  - 42,088 Medline Citations and abstracts from many life-science journals
  - 875 Patents Biology-related abstracts of patent applications
- 13 Ontologies
  - 13 GO Gene Ontology
  - 0 SBO Systems Biology Ontology
  - 0 Taxonomy NCBI Taxonomy database of Organism names

- 4 EBI Web Site
  - 4 Main sections
  - 0 EBI Members and Groups
  - 0 2can Support Portal

Tools and Services

The EBI (2007) front webpage is also a portal to many tools and services:

- EMBL-Bank
- UniProt
- ArrayExpress
- Ensembl
- InterPro
- PDB-EBI
- Genomes
- Nucleotide sequences
- Protein sequences
- Macromolecular structures
- Small molecules
- Gene expression
- Molecular interactions
- Reactions and pathways
- Protein families
- Enzymes
- Literature
- Taxonomy
- Ontologies
- Sequence similarity and analysis
- Pattern and motif searches
- Structure analysis
- Text mining
- Downloads

Interlinking of Tools and Databases

Many of the tools are highly interlinked, both to databases within the EBI (2007) and to Europe-wide and worldwide resources, for example via Ensembl (2007). Many of these tools can be accessed one by one. However, with the power of the bioinformatics grid from EMBRACE (2007), it is possible to develop workflows to combine resources across the EBI capabilities and across Europe. Thus collaborative research is also leading to collaborative data and service capabilities as well. The EMBRACE bioinformatics grid has published the *Web Services Development Guide* (EMBRACE-GUIDE 2007), which describes how to build EMBRACE-compliant Web services and access the full power of the grid being developed. There will also be a number of already developed applications on the project portal that are ready to use for the less specialist user. These capabilities promise to transform the way data is gathered, stored and analysed.

Distributed Development of Resources

Centres and Partners

A number of major resource centres in bioinformatics and systems biology exist around Europe, for example see the WTEK survey done in Europe (Cassman et al. 2007). An excellent list of institutions may be found by going through the lists of projects and looking at the partners and their individual websites. This involves hundreds of laboratories, and will provide listings of many of the key laboratories in Europe, and a clear indication of their activities. In the case of BioSapiens-partners (2007), and many others, the partners list corresponds to many of the major European institutions in bioinformatics, and also provides the e-mail contact addresses of key personnel

- European Molecular Biology Laboratory – European Bioinformatics Institute
- German National Centre for Health and Environment
- Universitè Libre de Bruxelles
- Consejo Superior de Investigaciones Cientificas
- Istitut Municipal d'Assisència Sanitària
- Genome Research Ltd.
- Max Planck Institute for Informatics
- The Hebrew University of Jerusalem
- Department of Biochemical Sciences University of Rome "La Sapienza"
- Stockholms Universitet
- Chancellor, Masters and Scholars of the University of Oxford
- University College London
- Stichting Katholieke Universiteit
- Swiss Institute of Bioinformatics
- Technical University of Denmark
- University of Helsinki
- University of Geneva
- Institute of Enzymology, Hungarian Academy of Sciences
- Universität zu Köln
- Institut Pasteur
- BioInfoBank Institute

- Max Planck Institute for Molecular Genetics
- Genoscope
- University of Bologna
- Centre de Regulació Genòmica
- The Centre for Research and Technology Hellas
- Fundación Centro Nacional de Investigaciones Oncológicas Carlos III
- Technische Universität Carolo-Wilhelmina zu Braunschweig

Collaborative Bioinformatics Resource and Infrastructure Projects

A number of other major bioinformatics and experimental projects have been supported in FP5 (2007) and FP6 (2007) to provide major and essential infrastructures in bioinformatics:

- TEMBLOR (2007) – major European Union gene/protein databases (see Chap. 2)
- FELICS (2007) – free European life science information and computational services
- Eurofungbase (2007) – European fungal genomic database

Database Infrastructure

FELICS (2007) is a major new infrastructure project to organise and make available a complete range of biomolecular information to life science research throughout Europe. It combines the work of the EBI (2007) and the Swiss Institute for Bioinformatics to create and provide public domain databases. The University of Cologne will put the enzyme information in the BRENDA (2007) database, a comprehensive enzyme information system, into the public domain as part of the interlinked collection of information. The European Patent Office (EPO 2007) will include the biomolecular information from its patents and databases (EPO-PatentSearch 2007). The 5-year project proposes a detailed collection of joint research activities, which develop and enhance the content of the databases and the connections between them. Connections to the wider community will be made through BioSapiens (2007) and EMBRACE (2007).

Fungal Database

In addition to projects to link and exploit the large genomic and proteomic databases, a number of specialist database projects are supported, such as Eurofungbase (2007), ensuring that the European biotechnology industries are well integrated with the research base in Europe. The objectives of the project are to develop the tools and technologies and databases to enable innovative functional genomic research of hyphal fungi. The project focuses on several filamentous fungi for different reasons. *Aspergillus nidulans* has a long record of use as a fungal model organism. *Aspergillus niger*, *Trichoderma reesei* and *Penicillium chrysogenum* are important cell factories used for the production of enzymes and metabolites, including compounds such as β-lactams, with benefits to human health. The human pathogen *Aspergillus fumigatus* not only serves as a model pathogen, but is also becoming more and more of a serious threat to human health.

Distributed Annotation System

Distributed Annotation

One of the main objectives of BioSapiens (2007) is to provide a large-scale, concentrated effort by laboratories distributed around Europe to annotate genome data, using both informatics tools and input from experimentalists. To integrate the available annotation, DAS (2007), based on developments by Dowell et al. (2001), is used as the core technology. DAS (2007) is a specification of a protocol for requesting and returning annotation data for, e.g., genomic regions, genes or protein sequences. The annotations are stored decentralised by third-party annotators, and are integrated on an as-needed basis by client-side software like DASTY (2007), SPICE (2007) or Ensembl (2007). To learn more about DAS (2007), how it can be used and how to set up a DAS server, see the Ensembl (2007) DAS help pages.

Annotation Server Information Service

The BioSapiens (2007) DAS Server Information Resource (BioSapiensDIR) lists all available information about the DAS (2007) servers within the BioSapiens (2007) network. This information is provided by DAS Registry. There are currently 70 distinct DAS (2007) servers providing 70 different data sources. The portal allows visitors to view protein annotations, genomic annotations and structural annotations; hence, it provides a major resource for the biomedical research community and the opportunity to participate in annotation activity, even if one is not a member of the BioSapiens (2007) project. The DAS (2007) portal is designed to provide a simple interface to utilise various DAS clients to view annotation, feature and structural information of proteins. The portal has two main components, an input area and a search area. Several DAS clients are available:

- Ensembl (2007) is a joint project between EBI (2007) and the Wellcome Trust Sanger Institute (WTSI 2007) to develop a software system which produces and maintains automatic annotation on metazoan genomes.

- DASTY (2007) is a Flash DAS client, which queries various protein DAS servers and visualises protein sequence features.
- SPICE (2007) allows visualisation of protein sequences, structures and their annotations, using the DAS (2007) protocol.

The portal has two main ways to retrieve information, firstly through direct input of a UniProt (2007) or Protein Data Bank (PDB 2007) or GENCODE (2007) accession number or a chromosome location and secondly via a simple search mechanism. Hence, it is fully compatible with more usual data access means.

Database Grid Integration

A Bioinformatics Grid for Europe

EMBRACE (2007) was initiated to allow data providers and tool builders to standardise their data access and software tools for a more user-friendly approach to databases, in the field of biological and biomolecular information. EMBRACE (2007) addresses the need for integration of data and analysis resources for biological and biomolecular information by developing the tools to implement a bioinformatics grid for data and computing. The many publicly available collections of biomolecular information do a reasonable job for a given domain. Software tools to organise and analyse this information are available both from the public domain and commercially. In principle, cross-references in these databases allow interdatabase navigation; however, the links are sparse and coarse-grained, and their exploitation requires biological knowledge and expert programming. As a result, every serious bioinformatics centre is burdened not only with the task of maintaining local data and software, but also of supporting users in the substantial task of exploring the natural biological connections between data. This requires considerable human effort. Current trends in systems biology demand greatly improved connections between different domains of knowledge, and the weaknesses in information integration are becoming an intolerable hindrance. EMBRACE (2007) is addressing these weaknesses by enabling data providers and tool builders to standardise their data access and software tools, using the new grid computing technologies that are ideally adapted to the task. The use of these standard methods allows data resources to be essentially self-describing, allowing software to work out the structure of the data, in large part automatically. Apart from facilitating widespread integration of software and data, this makes the interacting systems easy to update; for example, it reflects changes to the internal representation of the data.

Grid Connectivity and Worldwide Access

Grid capabilities for data and computing had already been established via European and national grid hardware and middleware networks (e.g. the GEANT2 2007 and

EGEE 2007 grid projects funded by European Commission Directorate General for the Information Society). These capabilities include links from the pan-European backbone to worldwide grid networks in:

- North America through the NASA, Abilene, ESnet and CA*net4 research networks
- Japan through SINET
- Southern and eastern Europe through SEEREN
- The Mediterranean through EUMEDCONNECT
- Latin America through ALICE
- The Asia-Pacific region under TEIN2

However, detailed Web software, Web services and protocols for information presented externally by bioinformatics databases needed to be established to make use of this capability. This is what is being accomplished by the EMBRACE (2007) grid.

Objectives of a Grid Infrastructure

The objective of EMBRACE (2007) is to draw together a wide group of experts throughout Europe who are involved in the use of information technology in the biomolecular sciences. The network will optimise informatics and information exploitation by pure and applied biological scientists, in both the academic and the commercial sectors. The result is highly integrated access to a broad range of biomolecular data and software packages. Groups in the network are involved in the following activities:

- Collection, curation and provision of biomolecular information
- Development of tools and programming interfaces to exploit that information
- Tracking and exploiting advances in information technology, with a view to applying them in bioinformatics training and also to reaching out to groups who can benefit from the work of the network

These groups work together to enable highly functional interactive access to a wide range of biomolecular data (sequence, structure, annotation, etc.), and tools with which to exploit the data. This naturally includes many core databases and tools available from the EBI (2007), but crucially the methods used will support the integration of dispersed, autonomous information. As a result, groups throughout Europe will be expected to integrate their own local or proprietary databases and tools into the collaborative "information space" which constitutes the EMBRACEgrid – a "data grid" allowing integrated exploitation of data, analogous to a "computational grid", which enables unified exploitation of dispersed computer resources. EMBRACEgrid will serve as a comprehensive virtual information source: virtual in the sense that it will have no single physical location, being rather a dispersed set of tightly coupled resources. EMBRACEgrid will be a permanent product of the project.

Expected Grid Results

The results expected are as follows:

- Standardised application programming interfaces (APIs) with all the core biological databases at the EBI, as well as with several wide-ranging sources of other information distributed throughout Europe
- Software tools that exploit the data through the new APIs, to provide a working environment in which to access and analyse the data, and also to facilitate the development of further tools in a consistent programming environment
- Technological standards for finding and describing the data and application services mentioned above
- Training and outreach to enable biologists to get the best out of the resulting tools and data, and bioinformaticians to develop ever-better tools, in the knowledge that they are firmly connected to all the data.

Linked Databases

There is currently a great deal of investment in postgenomics projects. About once a week the sequence of an entire species becomes available. Transcriptomics and proteomics projects are producing data avalanche after data avalanche. Each time (bio)informaticians have dealt with the data flow of one type of project, two new types of high-throughput experiment have been developed. All these data are finding their way to the biosciences, and fields such as the pharmaceutical industry, health, food and agriculture are all likely to undergo major revolutions that are expected to improve the quality of life for all, from infants to the elderly. This project sits in the context of existing integration projects such as Integr8 (2007) and BioMart (2007). These projects, and information resources like Ensembl (2007), exploit the standards developed in EMBRACE (2007) to provide common interfaces to data and tools across Europe, targeted to the needs of experimental research. Very significant progress has already been made, and the basis is now fully established for an integrated grid for bioinformatics and systems biology. The content being included involves a wide range of essential databases, described in EMBRACE (2007) work package 1:

- EMBL-Bank nucleotide sequence database
- Swiss-Prot, TrEMBL, UniProt
- Ensembl
- InterPro
- ELM
- SMART
- Macromolecular structure data
- ArrayExpress
- Literature

- Orthologues in all completely sequenced genomes
- Untranslated regions
- Three-dimensional electron microscopy data
- ProDom
- GenomeMatrix
- PairsDB
- CATH
- Gene3D
- ORFandDB
- Regulatory single-nucleotide polymorphismss with disease causing potential

Linked Tools

The tools being integrated in the grid include:

- EMBOSS
- SMART
- ELM
- PatSearch
- UTOPIA
- CINEMA
- NPS@
- modHMM
- Palign
- Modules for PDB files
- Homology modelling
- Gepardi
- NCUT
- Functional annotation

Open Software Linked Tools

Of these tools, some are huge toolboxes in their own right, e.g. EMBOSS (2007), the European Molecular Biology Open Software Suite. This package has been incorporated into the EMBRACE (2007) grid. The application of grid technologies to such major analysis packages will result in major improvements for all users. EMBOSS is a free open source software analysis package specially developed for the needs of the molecular biology user community (EMBNET 2007). The software automatically copes with data in a variety of formats and even allows transparent retrieval of sequence data from the Web. Also, as extensive libraries are provided with the package, it is a platform to allow other scientists to develop and release software in true open source spirit. EMBOSS also integrates a range of currently available packages and tools for sequence analysis into a seamless whole. Within EMBOSS there are around hundreds of programs (applications) covering areas such as:

- Sequence alignment
- Rapid database searching with sequence patterns
- Protein motif identification, including domain analysis
- Nucleotide sequence pattern analysis – for example to identify CpG islands or repeats
- Codon usage analysis for small genomes
- Rapid identification of sequence patterns in large-scale sequence sets
- Presentation tools for publication, and much more

Popular applications include:

- prophet – gapped alignment for profiles
- infoseq – displays some simple information about sequences
- water – Smith–Waterman local alignment
- pepstats – protein statistics
- showfeat – shows features of a sequence
- palindrome – looks for inverted repeats in a nucleotide sequence
- eprimer3 – picks PCR primers and hybridisation oligos
- profit – scans a sequence or database with a matrix or profile
- extractseq – extracts regions from a sequence
- marscan – finds MAR/SAR sites in nucleic sequences
- tfscan – scans DNA sequences for transcription factors
- patmatmotifs – compares a protein sequence to the PROSITE motif database
- showdb – displays information on the currently available databases
- wossname – finds programs by keywords in their one-line documentation
- abiview – reads ABI file and display the trace
- tranalign – aligns nucleic coding regions given the aligned proteins

Automated Gene Finding

The computational packages being developed each constitute a major research area in its own right, see EMBRACE-WP4 (2007). Perhaps the most interesting case for illustrating the power of grid applications is GeneFinder. It automates processes on which researchers might spend months of time in trying to accomplish similar results. Positional cloning of trait genes is very laborious and the amount of information on gene function is accumulating in different organisms so rapidly that it is hard for a research group to collect all relevant information without time-consuming searches in a number of databases. GeneFinder integrates different tools and databases by having a Web portal as a starting point, and evaluates the integration and content and graphical output of a given analysis. For example, a researcher has mapped a trait locus controlling a certain phenotype to a certain chromosomal region and would like to extract all available information on the genes in the corresponding interval and in all other species sharing a homologous chromosomal region. He/she would like to extract existing information on gene function, disease conditions, tissue expression, etc. to identify the most likely candidate gene for the trait. The resources normally used would be PubMed, Ensembl, Online Mendelian Inheritance in Man (OMIM), UNIGENE, a public microarray data repository, and Swiss-Prot. The researcher begins with a set of trait loci mapped to different chromosomal regions.

Manual Versus Automated Gene Finding

A specific example is a locus affecting muscle development in the pig that has been mapped to a chromosomal region that shares homology with a certain region in humans. Without the power of EMBRACE, the procedure would be as follows:

- Go to Ensembl and request all human genes that has been assigned to the homologous region in human (and possibly mouse and rat).
- Go to PubMed and search for articles about the genes in the actual region and try to find those genes which are expressed in muscle and have been reported to be involved with muscle development.
- Go to LocusLink and ask the same question.
- Go to OMIM and ask whether any gene in this interval has been associated with a muscle disorder.
- Go to the Mouse Genome Database (MGD) and ask the same question. Does any mutation in any of these genes give a phenotype in muscle?
- Go to UNIGENE and ask which of the genes in the actual chromosome region is express in muscle in humans, mouse, rat, etc.
- Go to all websites containing expression data and create an *in silico* expression profile. Return all genes that are expressed in muscle.
- Weight the number of positive results in the queries above and return a score to the user.

The expected end result would be a list of candidate genes for each trait locus including a score with the evidence supporting the gene as a likely candidate. Very extensive manual clicking, cutting and pasting would be involved. The EMBRACE solution is a bioinformatics portal that integrates several tools and databases to facilitate the finding of candidate genes located on a genomic region identified by positional cloning.

Sequence Motif Analysis

The PROSITE (2007) database contains a series of about 1,500 protein sequence motifs, many of which relate to post-translational modifications such as glycosylation andphosphorylation. Many motifs have a high probability of occurrence, i.e. the glycosylation signal N-{P}-S/T-{P} (or in words, an asparagine, followed by anything but proline, a serine or threonine, and again anything but proline) has a chance of about one in 200 of occurring in a random sequence. This implies that such glycosylation sites, when observed, have a high chance of being a false positive (i.e. being predicted as glycosylation site while actually not being one). Several simple methods can be used to reduce this number of false positives. For example, glycosylation has not been observed yet in transmembrane helices, and any post-translational modification motif is much more convincing when it is conserved among all (orthologous) members of that sequence family. EMBRACE (2007) will produce a bioinformatics server that will improve the PROSITE prediction of high-probability sequence motifs by filtering out motifs that fall in predicted transmembrane domains and by filtering out motifs that are not conserved in closely related members of the sequence family.

Integrating Promoter Motif Analysis and Gene Expression

Integrating promoter motif analysis and gene expression data represents a more complex system where the workflow will contain several bioinformatics tools and databases partially interacting with each other. Firstly, data need to be retrieved from a database and these data are then processed (clustering of microarray data). The result of this first process then needs to be compared with the content of a second database (search for regulatory elements) and the system will then, according to this comparison, decide on the next subset of data. The final dataset is then analysed by two different algorithms finding specific motifs. Finally, those motifs will be the input of the last bioinformatics tool delivering the data the user is looking for.

Protein Family Analyses

With use of ProDom (2007), a recurrent task was treated which is encountered in genome annotation and protein family curation: the inventory, domain analysis and classification of all proteins of a specific family encoded in a genome. Other more complex projects will include:

- Amplification of all genes pertaining to a domain family
- Phylogenetic analysis of all proteins containing a specific domain type
- Detection of novel protein types; detecting horizontal transfer events within a family; identifying phylum-specific protein families
- Domain-based three-dimensional modelling pipeline
- Identification of domains with putative novel folds for structural genomics

Ontologies

Gene Ontology

Ontologies now exist in many formats, such as Gene Ontology (GO 2007) at the genetics level and medical classifications at the physiological and pathophysiological levels The Gene Ontology Consortium (GOC) is an international collaboration among scientists at various biological databases, with an editorial office based at the EBI (2007). GO (2007) and GOA (2007) were both supported as part of the TEMBLOR (2007) project. A full introduction to the GO project, as well as links to search tools, data downloads and contact information, can be found on the GO homepage. The objective of GO is to provide controlled vocabularies for the description of the molecular function, biological process and cellular component of gene products. These terms are to be used as attributes of gene products by collaborating databases, facilitating uniform queries across them. The controlled vocabularies of terms are structured to allow both attribution and querying to be at different levels of granularity.

Gene Ontology Annotations

The GOA (2007) project provides high-quality GO annotations to proteins in the UniProt Knowledgebase (UniProtKB) and the International Protein Index (IPI) and is a central dataset for other major multispecies databases, such as Ensembl and those at the NCBI. In 2006 GOA became a central participant in the new GOC Reference Genome Annotation project and is committed to the comprehensive annotation of a set of disease-related proteins in human. With this project the GOC intends to generate a reliable set of GO annotations for the 12 selected genomes that will also empower comparative methods used in first pass annotation of other proteomes. Because of the multispecies nature of the UniProtKB, GOA also assists in the curation of other species. This involves electronic annotation and the integration of high-quality manual GO annotation from all GOC model organism groups and specialist groups (e.g. LifeDB). This effort ensures that the GOA dataset remains a key reference and a comprehensive source of GO annotation for all species.

Open Biomedical Ontologies

GO is itself a member of an international collaboration, the Open Biomedical Ontologies (OBO 2007). The range of ontologies is remarkable, with over 50 available, covering a wide range of function and model organisms.

Medical Ontologies

There are extensive ontologies available at the medical level as well as the genetic level, e.g. Disease Ontology (2007), a controlled medical vocabulary developed in collaboration with the Nugene (2007) project at Northwestern University. It was designed to facilitate the mapping of diseases and associated conditions to particular medical codes such as ICD-9-CM (2007), SNOMED (2007) and others. ICD-9-CM is based on the World Health Organisation's ICD-9. ICD-9-CM is the official system of assigning codes to diagnoses and procedures associated with hospital utilisation in the USA. Disease Ontology can also be used to associate model organism phenotypes to human disease as well as medical record mining.

Text Mining

Reference Searching Plus Full Data Overview

The standard world tool for reference searching is PubMed (2007). EBIMed (2007), which has improved functionality compared with PubMed, was developed as a Web application that combines information retrieval and extraction from MEDLINE. EBIMed finds MEDLINE abstracts in the same way PubMed does. Then it goes a step beyond and analyses them to offer a complete overview on associations between UniProt protein/gene names, GO annotations, drugs and species. The results are shown in a table that displays all the associations and links to the sentences that support them and to the original abstracts. Queries can be typed in the text box provided, while following the syntax conventions. Terms will be looked up throughout MEDLINE and several abstracts will thus be retrieved and analysed. In the simple interface the higher limit is 500 to make the process quick. A higher limit is available through the advanced search interface. By selecting relevant sentences and highlighting the biomedical terminology, EBIMed enhances the ability to acquire knowledge, relate facts, discover implications and, overall, have a good overview, economising the effort in reading. Indexed fields include PMID, AbstractText, ArticleTitle, AuthorList, MeshHeadingList, DateCreated, DateCompleted, DateRevised, PubDate, Language and MedlineJournalInfo.

Text Mining on Texts

Under FP5 (2007), the European Commission sponsored a number of text mining and annotation programmes, including eBioSci (2007) and BioMinT (2007). More recently, tools have been developed under FP6 (2007) which provide wide-ranging capabilities, such as iHop (2007), see (Hoffmann 2007). A network of concurring genes and proteins extends through the scientific literature touching on phenotypes, pathological conditions and gene function. iHOP provides this network as a natural way of accessing millions of PubMed abstracts. Using genes and proteins as hyperlinks between sentences and abstracts allows the information in PubMed to be converted into one navigable resource, bringing all the advantages of the Internet to scientific literature research. Text mining helps to define discrete function predictions of proteins, for example protein–protein interactions, protein phosphorylation sites and other proteins in the same pathway. This text mining information can be used in two ways: as a bona fide "function prediction" applicable to further analysis and as confirmed experiments to assess at least the sensitivity of various methods.

Whatizit (2007) is a text processing system that allows text mining tasks on text. The tasks come defined by the pipelines in a drop-down list and the text can be pasted in the text area. The description of each individual task/pipeline is available. Whatizit is also a MEDLINE abstracts retrieval/search engine. Instead of providing the text by copy and paste, a MEDLINE search can be launched. The abstracts that match the search criteria are retrieved and processed by a chosen pipeline. Whatizit identifies molecular biology terms and links them to publicly available databases. Identified terms are wrapped with XML tags that carry additional information, such as the primary keys to the databases where all the relevant information is kept. The wrapping XML is translated into HTML hypertext links. This service is highly appreciated by people who are reading literature and need to quickly find more information about a particular term, e.g. its UniProt ID. Whatizit is also available as a Web service and as a streamed servlet. The Web service allows enrichment of content within a website in a similar way to Wikipedia. The streamed servlet allows the processing of large amounts of text. In general, any vocabulary in the range of up to 500,000 terms can be easily integrated into a Whatizit pipeline. Whatizit is excellent at identifying formalised language patterns, and specialised, syntactically formalised, technical notation. Several vocabularies can be integrated in a single pipeline as is the case of Swiss-Prot and GO terms in the whatizitSwissprotGo pipeline. Examples of already integrated vocabularies are Swiss-Prot, GO, NCBI's taxonomy and Medline Plus.

CiteXplore (2007) combines literature search with text mining tools for biology. Search results are cross-referenced to EBI applications on the basis of publication identifiers. Links to full-text versions are provided where available. The help page explains which search options are available, how queries can be expanded and how to use wildcards. Text mining is provided internally by Whatizit and externally by iHOP.

Systems Biology Toolboxes

Systems Biology Toolboxes for Experimentalists

ENFIN (2007) will create the next generation of informatics resources for systems biology with a strong focus on the understanding of cell division. Despite the progress made in bioinformatics methods and databases to date, even the best experimental laboratories use only a small number of computational tools in their work, and they rarely exploit the potential of multiple datasets. The ENFIN network will transform the way computational analysis is used in the laboratory. The infrastructure will be entirely open, in the same way that genome information is accessible to all. To achieve its goals, the network will encourage close internal collaboration between experimental and computational research groups, and will have a specific consumables budget for testing predictions experimentally. The computational work includes the development of a core database infrastructure appropriate for the use by small laboratories, and the development of analysis methods, including Bayesian networks, metabolite flux modelling and correlations of protein modifications to pathways. The experimental techniques used to test this system include mass spectrometry, synthetic peptide biochemistry and RNA interference knockdown. Where appropriate, the network has chosen experimental areas related to intracellular signalling, associated with the cell cycle. The ENFIN network's specific objectives can be summarised as follows:

- Development of the ENFIN core (EnCORE), by taking a number of pre-existing database packages and providing a unified core platform which can be used in laboratories worldwide
- Curation of appropriate pathway knowledge and hypotheses
- Development and management of new experimental data standards and establishment of an annotated registry of databases
- Discrete function prediction
- Network reconstruction
- Systems level modelling (cycling between computational predictions and experimental validation feedback will be used to improve the accuracy of bioinformatics tools for predicting biological features)
- Critical assessment and integration, i.e. bringing together groups across the analysis layer, both to critically assess the methods and to uncover new synergies between computational and experimental groups
- Provision of graduate-level training, coordinated with the European School of Bioinformatics, so as to arrange short courses for graduate-level students
- Documentation from a Wet-laboratory perspective, enabling the consortium to develop a multiauthored resource, which is kept as current as possible through direct editing by appropriate researchers within the network
- Facilitating small and medium-sized enterprise (SME) outreach, to raise awareness of ENFIN in the biotechnology community, and to increase the understanding of SMEs' needs

Systems Biology Toolbox Applications Procedures

The main tangible benefit of ENFIN will be the development of a set of scientific procedures for understanding multicomponent systems using computational techniques via a systems biology toolbox. The ENFIN infrastructure will be available firstly to all the network's partners, and then to various identified collaborators and other interested laboratories. Remote access to the ENFIN core will allow on-site processing of information, access to public archived data as well as local information, access to specific, designed analysis methods which have been tested experimentally, and access to documentation written from the wet-laboratory perspective. ENFIN will have an impact on the understanding of complex human diseases such as cancer and diabetes. The systems-level vision of ENFIN is applicable to many other complex diseases and biological processes in other fields. As well as mammalian cells, systems biology could potentially impact the understanding of many pathogenic organisms – both eubacteria and eukaryotes. In addition to the network's research being fully integrated with that of leading molecular biology laboratories, it will organise a public, highly visible conference to explicitly test the outputs of systems biology predictions.

Yeast Systems Biology

The Yeast Systems Biology Network (YSBN 2007) project uses the yeast *Saccharomyces cerevisiae* as a model system, in order to advance the understanding of cellular systems. The central focus of YSBN is on facilitating cooperation between experimental and theoretical yeast researchers, thus exploiting the interdisciplinary characteristics of a systems biology approach. The YSBN project aims to provide a platform that will integrate data acquisition, data generation, modelling and recursive model optimisation. The achievement of the overall objectives of YSBN involves meeting the following targets:

- Offering a definition of standards for the documentation of proteome, transcriptome, metabolome, interactome, locasome and fluxome data
- Developing the structure of a database, focused on *Saccharomyces cerevisiae*, allowing for queries about experimental conditions and data from miscellaneous sources
- Defining the standards that will establish quality criteria for models to ensure sustainable model development
- Creating and maintaining software tools for mathematical cell modelling
- Organising an international conference, and creating a YSBN website that can function as a port, allowing the entire international community to access the tools produced by YSBN
- Establishing a platform for high-quality interdisciplinary student training in systems biology in Europe
- Illustrating how systems biology can be utilised in the design of efficient cell factories, for the production of fuels and chemicals

The project provides links (YSBN-tools 2007) to several systems biology tools, for example to the silicon cell modelling packages.

Expression Tools

In DIAMONDS (2007) deliverable D5.2, "Definition of the essential design characteristics of the toolbox", there is a description of the toolbox that is planned to be made available. Functionalities relating to expression data analysis will be implemented using Expression Profiler from EBI (2007). Expression Profiler will provide also a workbench for DIAMONDS, as well as direct access to the ArrayExpress (2007) public repository for microarray gene expression. In a second stage of platform implementation it will be expanded with additional functionalities, like:

- Detection of genes periodically expressed
- Promoter analysis
- Pathways annotation
- Modelling and simulation

In a third stage, more functionality will be added:

- For protein classification, Protonet and Pandora
- For gene homologies, blast processing, TREECON and i-ADHoRe

Power-Law Analysis Tools

The PLmaddon package at COSBICS (2007) was developed in the context of the Systems Biology Toolbox and MATLAB for analysis of power-law models. The module offers a set of functions for the analysis of S-system and GMA models, two kinds of power-law models used for mathematical modelling of biochemical systems. The current version of PLmaddon includes the following features:

- Conversion of models from matrix format to Systems Biology Toolbox format
- Estimation of steady states of the system in S-system and GMA models
- Local stability analysis
- Estimation of logarithmic gains
- Estimation of sensitivities
- Functions for improved visualisation of logarithmic gains and sensitivities

Additional power-law models functions are being implemented for the new version of the module, including:

- Local expansion of other biochemical models as power-law models
- Optimisation with biotechnological purposes
- Analysis of systems with moiety conservation

Since the package is fully integrated in Systems Biology Toolbox, the general purpose functions included in this package can also be used for the analysis of power-law models, including conversion to SBML (2007) format.

Multigenic Disease Modelling Platforms

The European Modelling Initiative Combating Complex Diseases (EMI-CD 2007) is directly aimed at medical applications. The analysis of the processes involved in the course of multigenic diseases necessitates coping with data from diverse experimental platforms. Consequently, important elements of the EMI-CD (2007) software platform are targeting data integration, as well as data standardisation. In particular, the EMI-CD (2007) platform is designed in a modular way. It main elements are:

- Database integration
- Experimental data integration
- Kinetic and probabilistic modelling of high-throughput data
- Modelling tool development
- Pathway annotation

Modelling Thousands of Reactions

The main purpose of EMI-CD (2007) is to provide a software platform complex enough to cope with various experimental techniques, aimed at discovering the gene function, and at understanding disease processes. A key goal is to cooperate with experimental projects on the design of experiments for combined strategies to combat human diseases (such as cancer and diabetes). Compatibility with other systems is also an issue, but by the use of SBML and more recently BioPAX, models can be interchanged between different systems. A further issue stems from scaling of the platform to large systems (i.e. whole-cell models). At the current stage, systems with a few thousand reactions are computationally feasible. EMI-CD (2007) is an open system for the integration of advanced analysis tools and other database systems.

Platform for Editing, Analysing and Varying Biochemical Models

In EMI-CD (2007) deliverable D1.1, "Open accessible Web-based interface for the population of object-oriented biological models", the use of PyBioS is discussed. PyBioS is designed for applications in systems biology and supports modelling and simulation. In contrast to, e.g., Gepasi or E-Cell, which are installed locally, PyBioS (2007) is a Web-based environment running on a server. The purpose of PyBioS (2007) is to provide a framework for conducting kinetic models of various sizes and levels of granularity. The tool can be used as a standalone modelling platform for editing, analysing and varying biochemical models in order to predict the time-dependent behaviour of the models. Alternatively, the platform offers the possibility of database interfaces (for example KEGG 2007; Reactome 2007; SRS 2007) where models can automatically be populated from database information. In particular, the high level of automation enables the analysis of large models. Predefined models can be selected from a model repository. Alternatively, the user can also create own models. Using the View tab, the user can inspect the hierarchical model. A list of all reactions of the model and a diagram of the whole reaction network are available via the Reactions and Network tabs, respectively. Model simulations performed by numerical integration of ordinary differential equation (ODE) systems are possible via the Simulation tab. The Population tab offers functions for the creation and modification of a model, e.g. forms to edit kinetic parameters and initial concentrations of model components representing state variables. A systems biologist might be interested in how the steady-state behaviour of a system might change if one parameter (e.g. the rate of glucose uptake of a cell) is varied. PyBioS (2007) provides functionalities to scan the steady-state behaviour (via successive simulations or root finding) given a varying parameter. This function, which also includes basic stability analysis, as well as other functions like the computation of conservation relations, metabolic control analysis, parameter fitting, etc. are provided via the Analysis tab. Finally, via the Export/Import tab, users can exchange models with other modelling systems, e.g. via SBML level 1 (see also EMI-CD 2007, deliverable D1.2). One important feature of PyBioS is the possibility of using information from public databases directly for the creation of models. It offers, e.g., an interface to the metabolic data of KEGG (2007) or an interface to the Reactome (2007) database. PyBioS (2007) acts as a model repository and supports the generation of large models based on publicly available information like the metabolic data of the KEGG database or Reactome. An ODE system of a model can be generated automatically based on predefined or user-defined kinetic laws and used for subsequent simulation of time-course series and further analysis of the dynamic behaviour of the underlying system. In the face of a lack of model parameters, large models are operated with random distributions of model parameters that are repeated multiple times and are validated by statistical arguments in order to generate reproducible model predictions. The forward-modelling approach supports the formulation of hypotheses, e.g. for *in silico* knockout experiments or time series.

The method developed in EMI-CD (2007) has contributed several unique elements that will ultimately strengthen systems biology approaches in general and disease-oriented research in particular. All developments provide user interfaces for practical applications:

- The modelling and simulation platform PyBioS (2007) capable of integrating multiple databases and experimental data with computational models
- Model analysis methods for kinetic models that support bottom-up and top-down approaches
- SRS (2007) system extended to multiple pathway resources supporting computational modelling
- Integrative database for transcriptome and proteomics primary data supporting computational modelling

- Model analysis methods for discrete probabilistic models integrated in the Metareg software system
- Reactome (2007) pathway database annotation of nine important disease-relevant pathways

Object Classes for Biological Entities from Gene to Cell

The underlying object-oriented structure of PyBioS entails a set of predefined object classes for biological entities (in the following referred to as BioObjects). Available BioObjects are Cell, Compartment, Compound, Chromosome, Polypeptide, Protein, Enzyme, Complex, Gene, etc. All of these BioObjects correspond to their respective biological counterparts and can be used for the creation of computational models that are hierarchically ordered according to fundamental cytological and molecular structures (e.g. compartments or molecule complexes). Object-specific information is stored as properties of the BioObjects, for instance each BioObject has an identifier and a concentration, and a Chromosome or Polypeptide can have a nucleotide or amino acid sequence, respectively. Furthermore, each BioObject can have one or more actions. An action, for instance, can be a chemical reaction or a transport process of molecules between different compartments. Actions describe the stoichiometry of the reactions and their kinetics. PyBioS provides a list of several predefined kinetic laws from which an appropriate one can be chosen for a specific reaction. Kinetic laws can be defined. The hierarchical object-oriented model composed of several BioObjects is internally stored in an object-oriented database and is used for further applications provided by PyBioS. For instance, for a time-course simulation, the object-oriented model is used for the automatic generation of a system of ODEs. To do actual modelling, in EMI-CD (2007) deliverable D2.2, "Library of kinetic models", several examples have been implemented, ranging from simple metabolic reactions to highly complex models, such as the cell cycle.

Stochastic Cell Simulation

The SmartCell (2007) cell network simulation program was further developed under COMBIO (2007) to model the evolution of a network in one cell. Based on stochastic algorithms, SmartCell needs multiple runs to have meaningful results. To help the user, SmartCell is distributed with a graphical user interface that allows the creation of models with a user-friendly interface, and also the analysis and treatment of results after the runs.

Chapter 6
Supporting Infrastructures

Abstract Analysis and modelling can only be useful when based on strong supporting resources and infrastructures. This chapter gives a quick survey of a few examples selected among larger facilities to demonstrate critical resources needed for bioinformatics and systems biology support. Good data are essential, generated by various wet-laboratory research infrastructure institutions. Needs for research infrastructures are discussed. Translation of research capabilities to health applications relies on better links with medical information systems. Overall knowledge management systems are necessary to integrate data and modelling practices. Perhaps the most important infrastructure is the education system, discussed here in the collaborative research context. Similarly, publishers, publications and their databases still constitute the ultimate knowledge base.

Introduction

Wet-Laboratory Infrastructure and Data

Bioinformatics research depends on the availability of high-quality, well-characterised data for its analyses, and systems biology even more so, since the data often need to be generated interactively and self-consistently with the analysis. There are many wet-laboratory research infrastructures around Europe, but only a few of the larger ones are discussed here to illustrate the types of resources needed and already available. A workshop was also held to discuss future priorities for biomedical research infrastructures, and the findings are discussed here. Although this book concentrates on basic biology databases, a huge amount of information is also held in medical information systems, which are very closely linked to patient data. The wide range of data and data types often need to be linked together into targeted knowledge management systems, and methods for doing this are discussed.

F.B. Marcus, *Bioinformatics and Systems Biology*,
DOI 10.1007/978-3-540-78353-4,

Education, Human Resources and Publications

Successful research is supported by more than facilities and data and tools. The basis of all research is education of new researchers and continuing education of established researchers. Some of the additional training programmes from collaborative research projects are discussed here. Dissemination of information and data has a critical role to play. Some of the roles that publishers might play are examined in the context of making data available. Other key resources are coordinating and other institutions involved in collaborative research projects. The direct collaboration usually involves a small fraction of their total resources, many of which are mobilised in support of collaborative efforts. The publications generated by these projects are major resources in and of themselves, and the consolidated publications of a project are often an excellent reference source for major fields of research.

Wet-Laboratory Research Infrastructures

European Collaborative Projects

An extremely extensive set of research infrastructure and resources is being provided collectively by FP6 (2007) research projects. Far too numerous to discuss here, they are accessible through the project catalogue FP6-Projects (2007). In this search website, selecting "Life Sciences" lists 367 projects, many of which provide key resources. The largest projects are further selected by choosing "Networks of Excellence" and "Integrated Projects", reducing the number to 128 projects. There are also major relevant projects under "Research Infrastructures" and "Information Society Technologies". It is also possible to find additional information about the FP6 projects discussed in this book.

A European Systems Biology Institute

The EMBL (2007) "Strategic Forward Look" document (EMBL-Strategic 2007) states that its conceptual focus for the next decade should shift to emphasise systems biology, an approach that EMBL started to utilise in 1997 in an exploratory manner and which is now a major focus. Systems biology is becoming the key EMBL-wide objective over the coming years, as the best route to understand biological modules and systems, using:

- Broad strength in cutting edge experimental science
- Past and present record in innovative computational science
- Tradition of interaction and integration across disciplines
- Turnover system, which is an ideal organisational framework with which to tackle new challenges like systems biology

Workshop on Research Infrastructures

A workshop was held in Brussels (Faure et al. 2005) on "Future Needs for Research Infrastructures in Biomedical Sciences" in Europe. This workshop provided an excellent survey of "wet-laboratory" and physics infrastructures that are currently available to support computational work in bioinformatics and systems biology, as well as the computational infrastructure itself and future requirements:

Model Organism Biobanks

Even the major EMMA (2007) mutant mouse archive facility cannot meet the full needs of the research community, so linked resources at widely separated geographical locations are necessary. A bottleneck is also disseminating results and resources. In the future, the resource centres should be linked to enlarge capacity, include all lines, add new types and partners, and link to phenotyping centres.

At the NASC (2007) European Arabidopsis Stock Centre, there are 500,000 accessions, with 30,000–40,000 being added per year. A major problem is gathering data and samples from the research community, of which there is an exponential increase. The centre has an important bioinformatics component for expression data analysis.

Protein Structure Facilities

The genomic revolution has opened the need and the challenge of characterising the products of the genome of a continuously increasing number of organisms. The knowledge of the structure of their proteome is a fundamental and necessary step for the understanding of protein function and of the processes in which they are involved.

X-ray Crystallography

Two main techniques are used for protein structure determination, X-ray crystallography and nuclear magnetic resonance (NMR) spectroscopy, and electron microscopy is also becoming important. X-ray crystallography, the first method to be developed and applied, is the most extensively used. EMBL Hamburg (EMBL-Hamburg Outstation 2007) is located on the site of DESY (German Synchrotron Research Centre), which provides synchrotron radiation through its DORIS positron storage ring. This radiation is used to study the structure and function of proteins using state-of-the-art equipment and methods. DESY was built for high energy physics studies, and the synchrotron radiation is a by-product from circulating electrons that is used for biological structures research. The bremsstrahlung (braking) radiation from the highly relativistic electrons is highly concentrated in a forward cone at a tangent to the storage ring, and thus is highly convenient for designing high-intensity beam lines. EMBL Hamburg operates seven synchrotron radiation beam lines, five of which are dedicated to biocrystallography, one to small-angle scattering from biological samples and one to X-ray absorption spectroscopy (extended X-ray absorption fine structure). The research developments are paralleled by an integrated approach to carry out scientifically demanding projects in structural biology, with facilities in molecular biology, heterologous expression in prokaryotic and eukaryotic hosts, protein purification, biophysical characterisation and crystallisation, complementing X-ray data acquisition and processing infrastructure. Scattered radiation patterns from the crystal structures can be inverse Fourier transformed to infer the three-dimensional atomic structure. Structure determination and interpretation is carried out on high-performance computers and state-of-the-art graphics facilities. In biology, a major principle is that protein function is directly dependent on structure. Currently 30,000 structures are available in worldwide databases, of which there are just 3,000–4,000 without redundancy. Of these, 80% have been determined by crystallography, and 90% using synchrotron radiation. There are 1,000 times more protein sequences than protein structures; hence, extrapolation algorithms are essential tools for inferring other structures.

NMR Spectroscopy

NMR is the other main technique to determine macromolecule structure. NMR spectroscopy allows the study of macromolecules of biological interest at atomic resolution in conditions similar to physiological ones, and allows direct interrogation of nuclei of specific atoms of the molecule and to achieve information, in addition to their position, i.e. the structure, also on how they move and on what surrounds them. Such information is used for the characterisation of a protein. The study of the relationship between the three-dimensional structure and the biological function of proteins, in physiological conditions, represents a field of primary importance. Furthermore, NMR is the most suitable technique to study protein–protein, protein–DNA/RNA and protein–small molecules interactions. Its power in this respect resides in the fact that most of the physiologically relevant interactions are weak, and the various partners exchange with free ones on fast time scales, thus preventing co-crystallisation. NMR is the key technique in this respect, as it is performed in solution and can detect even very weak interactions. It therefore has a tremendous importance and role in the pharmaceutical industry in various steps for drug design and screening. NMR has allowed the study of structure–function relationships of metal-coordinating proteins that are important for charge transfer processes, radical degradation processes and metal ion transport. The CERM (2007) Magnetic Resonance Centre is involved in postgenomic research in order to study proteins involved in important cellular processes. Calculation programs and mathematic algorithms are also used in CERM to study several biological systems.

CERM currently has nine NMR machines, with fields up to 22 T. There are multiple centres around Europe, with access is now provided on a voluntary basis.

Genomics

A major centre in Europe for genomics and genetics research is the Wellcome Trust Sanger Institute (WTSI 2007). The Sanger Institute has made major contributions to the sequencing of the genomes of humans, yeast, the nematode worm, the mouse and other model organisms, and more than 30 pathogens. It was a major partner in the international programme to map the mouse genome. It has also sequenced the entire zebrafish genome. The Sanger Institute generates around 60 million bases of raw sequence data daily. The rate of output has increased about fourfold every year. Sophisticated software and a large investment in computer resources keep the data organised and underpin efforts to identify genes and other sequence features. Results are analysed and presented to end users through tools such as Ensembl (2007). The development of high-throughput tools has opened up new opportunities to explore gene function on a grand scale. Rather than study genes one at a time, these tools enable researchers to track the activity of thousands in a single experiment. Two key techniques in use are DNA microarrays and gene expression atlases, revealing which genes are active in living tissues. The complete human genome sequence was derived from several anonymous individuals, 72% of which is from a single male. Genetically, humans are 99.9% identical, but differ in minute detail, and these differences can be medically important. A key aspect is to identify sequence variation in human populations, and how specific variants contribute to health and disease, by identifying genetic variations such as single-nucleotide polymorphisms (SNPs) and patterns of SNPs that are inherited together (haplotypes). A major initiative, the Cancer Genome Project (2007), is systematically searching all human genes for genetic variations implicated in cancer – a quest that has already unearthed new cancer genes and potential new therapies. Other new projects range from addressing biomedical questions, such as how chromosomes rearrange and evolve, to uncovering the molecular basis of disease from deafness to diabetes.

Proteomics

Proteomics is a broad and rapidly growing field. After the sequencing of the human genome, it is now clear that much of the complexity of the human body resides at the level of proteins rather than at that of the DNA sequences. This view is supported by the unexpectedly low number of human genes, and the high estimated number of proteins (of order one million) generated from these genes. For example, it is estimated that on average human proteins exist as ten to 15 different post-transitionally modified forms, all of which presumably have different functions. Much of the information processing in healthy and diseased human cells can only be studied at the protein level, and there is increasing evidence to link minor changes in expression of some of these modifications with specific diseases. Together with rapidly improving technology for characterising the proteome, there is now a unique chance to make an impact in this area. Using cutting edge technology, the Center for Proteome Analysis (CPA 2007) can separate cell or tissue samples into more than 17,000 proteins. The resulting gel images are then quantitated accurately using image analysis software (written and developed in-house) to detect the subtle changes in levels of expression related, for example, to the development of a disease. This high resolution provides excellent separation of post-translational modifications, which is of critical importance since particular modifications often play decisive roles in disease development. These highly purified proteins (or particular post-translational modifications) can then be recovered from the gels, in amounts, although small, sufficient for their identification by mass spectrometry, or to raise antibodies. The application of proteome analysis using two-dimensional gel electrophoresis has allowed CPA to make discoveries with great potential relevance for the treatment of a number of the major diseases which afflict humanity, including diabetes, cardiovascular disorders (ischaemia, hypertension, heart transplantation), cancer (cervical, breast and colon), cancer metastasis, rheumatoid arthritis and neurological diseases (Parkinson's disease and epilepsy) and ageing.

Imaging Living Systems

Imaging living systems is essential for understanding disease mechanisms, development of methods of non-invasive disease monitoring, drug discovery and assessment of therapeutic efficacy. Research is performed at imaging facilities at the microscopic, cellular, functional, structural and systems levels. The clinical neurosciences are dealing with diseases of increasing relevance to our societies of the future: dementia, stroke, movement disorders, psychiatric disease and behavioural disorders of the elderly. One approach to the investigation of these disorders that has become influential and increasingly useful is brain imaging; both functional and structural. There is already good evidence that certain neurodegenerative diseases can be detected with accuracy and sensitivity in the *preclinical* phase when degeneration is already considerable but clinical symptoms are not yet manifest because of cerebral compensation mechanisms. The opportunities for preventive or effective delaying treatments and measurement of their efficacy then becomes a very interesting proposition. The hardware needed for optimal imaging is very expensive and improved performance (for example by higher-power magnets for magnetic resonance imaging (MRI) scanning) demands continuing research and development funded by large investments in money and expertise. The Functional Imaging Laboratory (FIL 2007) undertakes research into the functional anatomy of the human brain in health and disease, using positron emission tomography (PET) and functional MRI (fMRI) techniques. The research programme divides into groupings by subject matter, but is essentially multidisciplinary. Thus, imaging is a key method for medical research, and is important to society; industry, medicine, and brain research. New technologies are coming on board, as are new methods, analytical tools, principled methods of data interpretation and applications. The field is diverse and is growing very quickly.

Three-Dimensional Electron Microscopy

The aim of 3D-EM (2007) is to make Europe the leader in three-dimensional electron microscopy analysis. The integration of the leading European laboratories in electron microscopy should aid the development of standardised procedures and innovative equipment for comprehensive structural analysis based on advances in electron microscopy. These will be made accessible to the biological and medical communities via the creation of state-of-the-art centres with adequate regional distribution to provide access to instrumentation and protocols developed within the network.

Research Infrastructures: Future Priorities

The infrastructures workshop (Faure et al. 2005) identified future infrastructure needs by analysing the appropriate size and nature of infrastructures for responding to these needs, and thereby provided guidance on how to implement the FP7 (2007) goals for life sciences research. The workshop was structured around seven general types of research infrastructures' requirements relevant to biomedical sciences:

- Biobanks, living organism resources, molecular tools and reagents
- Protein structure facilities
- High-throughput genome, genotype and proteome phenotype, sequence and interaction measurement facilities
- Bioinformatics, databases, software, resources and grid-linked services
- Imaging living systems capabilities
- Ion and radiation therapy research facilities
- Clinical research infrastructures

European Strategy Forum on Research Infrastructures

Many of the recommendations in the workshop were taken up in the ESFRI (2007) recommendations, where the first roadmap was published in 2005. This identified 35 large-scale European infrastructure projects that were prioritised for selection following an independent European consultation and peer-review process.

Future Infrastructures for Bioinformatics

The ELIXIR (2007) proposal, European Life Sciences Infrastructure for Biological Information, proposes the construction and operation of sustainable infrastructure for biological information in Europe to support life science research and its translation to medicine and the environment, bioindustries and society. Such infrastructure should comprise:

- An interlinked collection of "core" and specialised biological data resources and literature
- Standards and ontologies for newly emerging data
- A major upgrade for the core information resources
- New data resources as appropriate
- Integration and interoperability of diverse heterogeneous data
- Rapid search and access through friendly portal(s) supported by appropriate computer hardware infrastructure
- Infrastructure linking core data resources and national bioinformatics data and service providers
- Infrastructure to enable distributed annotations and tool development
- The opportunity to establish infrastructures for life science information in the accession states
- Links between molecular resources and developing resources for medicine (e.g. biobanks), agriculture and the environment (e.g. biodiversity)
- Access to high-performance computing, through links to Europe's supercomputer centres
- Coordination and provision of training and outreach across Europe to enhance national efforts
- Strong links to European bioindustries to ensure the optimal translation of life science research into the bioindustrial sector in Europe

Bioinformatics Infrastructures for Systems Biology

A recent workshop was held (Cassmann and Brunak 2007) on infrastructure needs for systems biology. The workshop website also contains slides of the presentations and an excellent set of references. The main conclusions may be summarised as follows:

1. General issues within systems biology
   - Support specific projects, e.g. ontologies and databases
   - Establish panels for assessment of research projects
   - Conduct systems biology benchmark studies possibly including funding for pre- and post-prediction experimental data generation, evaluation and creation of standards
   - Generate procedures for research infrastructures

2. Experimental tools
   - Support the creation of common experimental protocols, selection of truly, validated, common cell types, tools for single-cell analysis, globally useful reagents, reporter constructs, etc., making experimental data more valuable for modelling within systems biology
   - Complement the ENCODE project by a large-scale proteomics effort (alternative modifications, localisation, structure, etc.)
   - Application of established experimental techniques in the context of systems biology
3. Databases
   - Establish criteria for long-term support for systems biology relevant databases
   - Support the development of standard representations enabling interoperability between databases and tools
   - Support data capture incorporating minimal information, using standard formats and semantics
   - Support and broaden BioMart-like data integration schemes going beyond sequence-centric approaches
   - Promote access to full-length paper text and repositories and promote semantic enrichment efforts
   - Support "workflow" schemes in the context of systems biology
4. Systems biology software, software repositories and modelling approaches
   - Support initiatives in multiscale modelling spanning molecular to multitissue organism levels
   - Support initiatives to make existing software development platforms interoperable using best practice approaches and standards
   - Initiate an infrastructure-related software support mechanism
   - Support systems biology software repositories which incorporate software curation
   - Support education in the use of software within systems biology
5. Systems biology training
   - Support education in the use of software within systems biology
   - Initiate collaboration on establishing curricula in systems biology
   - Support activities similar to competitions like iGEM (2007)
6. Priorities
   - Create a mechanism to support ongoing benchmark efforts with special emphasis on data generation
   - Start an effort on standards and interoperability for both databases, software and experimental systems
   - Exchange information on training programmes
   - Consult the Society for Systems Biology

Medical Information Systems

Medical Informatics

Medical informatics is a major area in its own right. Websites such as that of the UK Health Informatics Society, until recently the British Medical Informatics Society (BMIS 2007), with its related link sites, show the breadth of the field. In Europe, Infosoc (2007), the Directorate General for the Information Society of the European Commission, supports collaborative research and resources in this area via its eHealth programme though programmes such as INFOBIOMED (2007). INFOBIOMED aims to structure European biomedical informatics to support individualised healthcare. Its programme shows the emphasis on mixing medical information with genetic information, as follows:

- Support – a comprehensive state-of-the-art review on methods and tools relevant for the biomedical informatics field. It comprises methods, technologies and systems for data analysis, information retrieval and decision support.
- DiseaseCard – an information retrieval tool for accessing and integrating genetic and medical information for health applications. Resorting to this integrated environment, clinicians are able to access and relate data on diseases already available on the Internet, scattered along multiple databases.
- MIND – Microarray Information Database – a free Web-based database that allows the storage of all data retrieved from microarray experiments and analysis.
- OSIRIS – a search tool for articles that refer to the SNPs identified for a gene of interest. The large number of SNPs reported for many genes and, more importantly, the variability in gene and SNP nomenclatures make it difficult to find a convenient search strategy. OSIRIS is a search tool for the retrieval of articles from MEDLINE related to the genetic polymorphisms reported for a human gene. The variations considered are single-nucleotide polymorphisms, insertion/deletion polymorphisms (indel), microsatellites and named variations (e.g. Alu sequences).
- Biomedical Informatics state of the art – the project has developed descriptions of a broad state of the art and future challenges on biomedical informatics methods, technologies and tools, which have been implemented into Wiki pages and can thus be easily accessed (INFOBIOMED-Wiki 2007).

Knowledge Management Systems

Choices of Systems Complexity

A feature of the resources described in this book is that they are modular, to be able to range from the needs of an individual researcher up to large projects for research or drug development. Some examples follow.

Individual Researcher or Research Team in One Laboratory

Researchers may want to collaborate directly with others, or just access the data and tools available. The websites discussed in this book give a wealth of contact details. The collaborative projects are open to informal collaboration with outside researchers or teams. Many of the collaborations are with other groups worldwide. Most projects and institutes have public Web portals that allow access to the databases and tools discussed, once they have reached a mature stage. Via these tools, it is possible both to make use of the data and to participate and contribute data as well. The DAS (2007) system of BioSapiens is an excellent example of this. DAS can be used to view data via Ensembl. It is also possible to download DAS software and for a scientist to therefore contribute his/her own annotations. There are also more powerful tools available. EB-eye (2007) can be the first port of call of life science researchers. The test cases of the EMBRACE bioinformatics grid such as Genefinder provide powerful capabilities which can automate researcher's work. The EMBRACE Web services allow the researcher to customise access to data and computation resources, as in EMBRACE-Biomed (2007). The TAVERNA (2007) software tool is available to help bioinformaticians design software in a rapidly changing data environment.

Laboratory Large Project

Many laboratories have their own bioinformatics group for support and research, often tied in with local laboratory information management systems . These groups usually have a number of local resources plus links to remote facilities. The new collaborative projects open broad new opportunities to link into the major collaborative and linked networks, and make full use of resources. Major databases can be linked, and specialised interfaces can be designed that reflect the precise needs of the local research programmes.

Collaborative Research Teams at Multiple Multinational Sites

This book has attempted to show all the aspects of collaborative research programmes funded by the European Commission. In addition, it provides a number of successfully working examples for other funding agencies or organisations that may wish to set up similar collaborations.

Major Collaborative Research and Industrial Development Programme

This represents perhaps the greatest challenge for collaborative research, which is to link up very large numbers of databases and tools to support a broad-based, multiarea research programme in the most effective way. One example of the broad requirements for such a collaborative knowledge management system is indicated in the IMI (2007) IMI-Research-Agenda (2007) dealing with knowledge management. Key goals include support for the safety and the efficacy projects. Many of the resources discussed there have been addressed to a greater or lesser degree by the various projects discussed in this book. An example is the report on human genetic variation linked databases (Marcus and Mulligan 2006), of which many recommendations will be implemented by the FP7 (2007) project GEN2PHEN (see Chap. 9). Methods are discussed by which large numbers and wide varieties of databases can be linked and used.

Education

School of Bioinformatics

Many of the large collaborative projects give special emphasis to education programmes and outreach activities. One of the most successful and visible of these is provided by BioSapiens (2007), with the establishment of the European School of Bioinformatics (ESB 2007), to train bioinformaticians and to encourage best practice in the exploitation of genome annotation data for biologists. The courses and meetings are open to all scientists throughout Europe, and are available at all levels, from basic courses for experimentalists to more advanced training for experts. The projects EMBRACE (2007) and ENFIN (2007) also participate jointly with this European school, to give it an extremely wide range of activities and subjects taught. There is a clear need to train and recruit creative and innovative young scientists in bioinformatics, and at the same time to help users in experimental laboratories to keep up with the developments in the field. The European School of Bioinformatics provides extensive training at all levels, from basic courses for experimentalists to more advanced training for experts. The Permanent European School of Bioinformatics is intensive and has a substantial practical component. Its objective is to give a basic overview of the methods and tools available to the community. Training is critical to the success of any scientific network, and BioSapiens (2007) has a work package devoted to this essential element. This involves a commitment to offer over 400 researchers the opportunity to get acquainted with bioinformatics by holding ten 5-day introductory bioinformatics courses over the course of the 5-year grant period. Postdoctoral workers from within the network are instructors at the School, and are therefore offered a chance to enhance both their scientific and their teaching skills. Each of the schools was followed by a 3-day advanced workshop and known as the Permanent European School of Bioinformatics. Several ESB-Schools (2007a–g) courses were held. The school is held twice a year, in a different country every time, and consists of a 6 days at entry level for inexperienced users and newcomers to the field. It is followed, in the same location, by a workshop on one of the topics of interest of the network. ENFIN (2007) also participated in the fourth, fifth, sixth and seventh schools, and has details on its website.

Training Workshops and Courses

In addition to the European School of Bioinformatics, individual workshops are organised for members of BioSapiens (2007) in order to continue scientific training for those members working in the areas of science supported by this undertaking. A proportion of the BioSapiens training effort is being directed towards trainers of master's degree students across Europe. An initial meeting brought several of these trainers together to discuss ways in which experiences in this area of training could be mutually beneficial. Many other collaborative research programmes have similar types of workshops.

In order to make sure that people can use the products being developed, EMBRACE (2007) has established a number of training courses:

- Grid technology
- Data modelling
- Portal integration
- Grid and Web services
- Bioinformatics immunology
- Regulatory sequence motif
- Grid technology, principles
- Grid-based Web services
- Applied gene ontology
- Workshop on Bioclipse
- Modular protein architecture
- Sequence annotation
- DNA sequence assembly
- Membrane protein
- Protein domain analysis
- System biology
- Protein structure families
- Genome annotation

Outreach

Another vital aspect of education is outreach. In the BioSapiens (2007) outreach work package, the objectives are to:

- Improve visibility of bioinformatics in Europe
- Increase impact of computational methods in biology
- Explore the potential for collaboration with experimental biologist
- Foster worldwide interaction between computational biologists
- Reach society

Training Workshops and Courses

Part III
Disease-Related Collaborative Research and Computational Biology

Chapter 7
Infectious and Major Diseases

Abstract Bioinformatics and systems biology research approaches are leading to a number of important insights and routes towards treatments in diseases involving viral and bacterial pathogens, cardiovascular–pulmonary diseases, diabetes and neurological disease. Cancer, with its special genetic characteristics, is discussed separately in Chap. 8. The immune system is examined to understand in role in both healing processes and disease. Many aspects of the research are oriented towards enabling drug development applications. Major new initiatives are planned for the Seventh Framework Programme.

Introduction

Approach to Disease Research

Along with major national, charitable and international programmes, the European Commission within FP6 (2007) has a separate and sizable programme devoted to major diseases (Vanvossel 2005), with a large number of projects. This chapter concentrates on both bioinformatics and systems biology approaches to diseases. A summary of bioinformatics methods applicable to disease research has been made by Lengauer (2007), who participates in BioSapiens (2007). By communication and links within and between projects, cross-project synergies are produced that lead to a highly inter-related approach to these diseases. These projects are effective in translating basic research knowledge into direct applications to health, providing input and resources for more directly targeted health research programmes.

Computational Biology and Disease

The range of diseases is extensive (Kumar and Clark 2002) and complicated at both the physiological/pathological level (Underwood 2002) and the cellular level (Woolf 2000). Modelling and description of physiological processes has a long

F.B. Marcus, *Bioinformatics and Systems Biology*,

DOI 10.1007/978-3-540-78353-4,

tradition at the organ level (Vander et al. 2001) and the cellular level (Fall et al. 2002). Bioinformatics and systems biology provides the means to address these many disease areas. Important advances have been made in combining organ-level modelling with cell-level modelling in particular areas, for example in cardiovascular disease (Bock and Goode 2002). The diseases discussed here range from diabetes and diseases of the endocrine system (Laycock and Wise 1996), to diseases of several other bodily systems. Despite significant increases in pharmaceutical research and development spending, the number of new approved medicines has remained fairly constant. One possible reason is that analytical methods and tools are not yet fully installed in the drug development process. While bioinformatics is used in drug target discovery as discussed in Lengauer (2007) or for reactions to drugs (Nagl 2006), this is not the case for the later stages. In particular, the simulation and modelling of biological processes, such as disease-relevant signalling pathways and metabolic processes, are underdeveloped in drug target validation. *In silico* experiments could be the basis for successful screening, and the entire drug development process could be accompanied by bioinformatics and systems biology approaches. Multiple databases exist already, and a variety of experimental techniques have produced gene and proteome expression data from various tissues and samples, and important disease-relevant pathways have been investigated.

Viral and Bacterial Pathogens

Bacteria (Singleton 2004) and viruses (Perry et al. 2002) have had billions of years to develop, and include variants that are involved in a wide range of infectious and dangerous diseases. Major advances have been made in the molecular genetics of bacteria (Dale and Park 2004) and viruses, but there is still a long way to go to make the best use of this knowledge.

Cardiovascular–Pulmonary Disease

Chronic diseases are usually the result of interactions between individual susceptibility and different environmental and/or life-style factors, and are often modulated by multiple genes. The interplay between these factors determines disease phenotype and, hence, the prognostic and therapeutic implications of the disease. This interplay between genetically predetermined susceptibility and disease phenotype can in turn be revealed by computer analysis integrating clinical and biomedical data. Computer analysis related to clinical problems is currently in a phase of accelerated growth. Some examples of the application of computer analysis to clinical practice are the analysis of myocardial perfusion images and cardiograms and the development of a screening device for the diagnosis of heart murmurs. However, all the approaches currently implemented in clinical practice use very limited datasets, despite the availability of extensive data from the "-omics" revolution. Only by integrating genomic, proteomic and metabolomic data can knowledge that is useful for the understanding and treatment of complex human diseases begin to be obtained.

Diabetes

Bioinformatics methods for diagnostic screening are a bottleneck in current biomedical research. While exploratory methods – such as statistical hypotheses testing, clustering of gene expression profiles and classification methods – have been successful in the detection of molecular markers for interesting diseases, these techniques fail to validate these markers in their gene regulatory context and to integrate other data sources relevant for diagnostic purposes. For these tasks, novel modelling techniques, network analyses and data integration methods are indispensable. The analysis of processes involved in the course of complex polygenic diseases, such as obesity and type-2 diabetes, is in fact a multistep procedure that has to cope with data from diverse experimental functional genomics platforms (gene and protein expression), physiological data, environmental factors and others.

Neurological Diseases

Understanding and modelling neurodegenerative diseases (Alzheimer's, Parkinson's, etc.) will help to identify new drugs to cure these diseases and to design new therapeutic strategies. Epileptic seizures can happen to anyone at any time and one in 20 people will have an epileptic seizure at some time in their life. Diagnosis is made after one or more seizures and antiepileptic medication is generally given. The problem with the existing medication is that more than 50% of patients are resistant to the treatments (they keep on having seizures) and current treatments have important cognitive side effects (impairment of awakeness, loss of memory).

Immunology

The study of infectious diseases is tied up with that of the immune system (Roitt et al. 2001), one of the most complex systems in the body. Immune responses are mediated by a variety of cells, and by the soluble molecules which they secrete. Although the leucocytes are central to all immune responses, other cells also participate by signalling.

Drug Development References

Drug development (Sneader 2005) has a long history. The design of drugs (Smith 2006) has become a major modern discipline and industry (Klefenz 2002). With the advent of full genome studies, the research paradigms have been changing (Bock et al. 2000), but are only beginning to fulfil their promise and improve on the vast range of currently available drugs (Shannon et al. 2004). There are still major advances to be made in discovering and testing the drugs themselves (Rang et al. 2002) as well as biomarkers (Pagana and Pagana 2002) to indicate the diseases needing treatment and in the use of bioinformatics (Lengauer 2007; Nagl 2006; Bertau et al. 2007).

Viral Pathogens

HIV/AIDS and Hepatitis C Virus/Hepatitis C Coordinated Studies and Databases

BioSapiens-WP15 (2007) has made important advances in studying infectious diseases and their associated pathogens: HIV/AIDS and hepatitis C virus (HCV)/ hepatitis C. A wide variety of bioinformatics methods were applied, together with experimental data and know-how, in order to advance the understanding of the diseases, provide tools to optimise existing therapies and help in the development of new therapies and vaccines. The genes and gene products of the infectious agents were analysed to create a publicly available resource for collecting existing data and data produced. Major results of this work were summarised at a joint conference between BioSapiens (2007) and ViRgil (2007a, b) in BioSapiens-WP15 (2007) deliverable Del 15.1, "Conference report – BioSapiens-ViRgil workshop on bioinformatics for viral infections held in Bonn in 2005". ViRgil (2007) is a dedicated to combating viral drug resistance, and influenza and hepatitis B and C are in the focus of attention. The conference results are found on the website ViRgil-(2007a). There were discussions about use of the software platform ViralDAS (2007), which accesses databases on HIV1-HXB2 and HCV-1B and the euHCVdb (2007) European Hepatitis C Virus Database.

Bioinformatics Prediction of HIV Antiretroviral Resistance Trajectory

Important summaries of techniques used to predict optimum AIDS/HIV therapies are found at MPI-INF-Bioinformatics-for-HIV (2007) and BioSapiens-WP15 (2007). In particular, in Altmann et al. (2007), it was shown that the outcome of antiretroviral combination therapy depends on many factors involving host, virus and drugs. They investigated prediction of treatment response from the applied drug combination and the genetic constellation of the virus population at baseline. The virus's evolutionary potential for escaping from drug pressure was explored as an additional predictor. Different encodings of the viral genotype and antiretroviral regimen were compared, including phenotypic and evolutionary information, namely predicted phenotypic drug resistance, activity of the regimen estimated from sequence space search, the genetic barrier to drug resistance and the genetic progression score. The benefit of phenotypic information in predicting virological response was confirmed by using predicted fold changes in drug susceptibility. Moreover, genetic barrier and predicted phenotypic drug resistance were found to be the best encodings across all datasets and statistical learning methods examined. A prototypical implementation of the best performing approach is freely available for research purposes at Geno2pheno (2007), with the basic principles described by Beerenwinkel et al. (2001).

HIV Genotypes

EMBRACE (2007) is studying HIV to develop novel data mining methods for correlating specific HIV genotypes with drug resistance. The rapid replication of HIV-1 coupled with the introduction and selection of sequence polymorphisms gives rise to quasi-species of HIV-1 within an infected individual and a global distribution of viruses that are genetically divergent. HIV-1 genetic diversity has been stratified and a subtype nomenclature based on sequence variation has been described. There are 12 subtypes (A–H, J–K, N, O) and a variety of circulating recombinant forms of HIV-1 in the current schema. Profile and hidden Markov model based methods have been developed for subtyping HIV-1 protease and reverse transcriptase sequences and neural networks and support vector machines are being explored for linking these subtypes to resistance for specific reverse transcription and protease inhibitors and drug regimes. Correlations are being examined between structural changes in regions remote from the active site of aspartyl protease with drug resistance. Microarray technology is used to investigate host–pathogen interactions in HIV and other viral systems, e.g. herpes viruses. Novel clustering techniques have been developed for identifying co-expressed genes. Dynamic Bayesian networks are also being explored for modelling gene interaction networks. An integrated resource, BioMap (2007), is being developed which links protein family data and functional annotations with expression data to facilitate data mining by using prior knowledge of putative protein functions and interactions. Under existing solutions, current algorithms are restricted to searches on locally established protein sequence and function resources such as the VIDA (2007) virus database of homologous protein families and a local implementation of GO (2007) for viruses. The CATH (2007) protein structure classification database is being used to access relevant structural data and the Gene3D (2007) database of protein families is providing information on appropriate host families. Access to external databases containing relevant information is very time consuming and inefficient. The solution is access to more extensive data sources integrated by EMBRACE (2007) and improved mechanisms of data transfer from integrated resources via bioinformatics grid technologies. Access to these data will improve sequence and structure analysis for understanding functional regions in the viral genome in the context of coordinated host and viral gene expression. It will also assist analysis of molecular interactions between the enzymes and analysis of functional pathways involved in host–pathogen interactions. Improved access to more efficient search and analysis algorithms via EMBRACE (2007) partners and grid-enabled technology will also considerably improve analysis of host–virus interactions and drug resistance.

HCV Sequence Alignment

EMBRACE (2007) is examining how to build HCV sequence alignment via a workflow where a user can upload a set of sequences or retrieve sequences from a database followed by a multiple alignment process with graphical output for visualization.

Herpes

EMBRACE (2007) is studying herpes, B-cell response and maturation. The user starts with a consensus clustering and searches for co-regulated genes. Protein sequences are retrieved for these genes, and BioMap (2007) is used to map functionality to those sequences. Further services are required to search for homologous sequences and third-party annotation to achieve as complete an annotation level as possible. This will facilitate the user's selection of candidate genes for experimental investigation.

Bacterial Pathogens

Pathogens Bacillus anthracis and Staphylococcus aureus, Using Bacillus subtilis

BaSysBio (2007) will use the model bacterium *Bacillus subtilis* to gain insight into the global structure of the regulatory networks that control bacterial metabolism. The project will validate the general applicability of the findings, and integrate the modelling/experimental strategy developed in the highly tractable *B. subtilis* model, towards an understanding of regulatory networks controlling pathogenesis in disease-causing bacteria. BaSysBio is making several technological contributions:

- *B. subtilis* living cell arrays to study the temporal regulation and the design principles of the transcription networks that control the timing of gene expression
- Efficient chromosomal engineering techniques for Gram-positive bacteria, including the pathogens *B. anthracis* and *Staphylococcus aureus*
- Parallel flux analysis based on ^{13}C-labelling experiments in microtiter plates
- Adaptation and improvement of existing high-throughput technologies for the specific project needs

The conceptual aspect of BaSysBio is the development of a theoretical framework for comprehensive, system-wide data interpretation. This differs from the current focus of much of systems biology, which concentrates on signalling networks and metabolic networks reconstructed through comparative genomics. It extends conceptually beyond data acquisition and interpretation approaches, through quantitative interpretation with the mathematical rigor of computational models. By integrating the multiple regulatory levels in a biological system, models will have high potential to simulate them accurately, to predict novel systems properties and properties of uncharacterised systems components, and to drive mechanistic understanding of the global regulation of *B. subtilis* metabolism, and by applying these methods to pathogens, of the adaptive transcriptional responses to stresses encountered by cells during pathogenesis.

Bacillus Cell Factory

BACELL-HEALTH (2007) is studying *B. subtilis*, *B. anthracis*, *B. cereus*, *Listeria monocytogenes*, *Staphylococcus aureus* and *Streptococcus pneumoniae.* While bacteria remain one of the main threats to human health and well-being, particularly in the light of increasing resistance to antimicrobials, they also have the capacity to provide products (e.g. antibiotics, vaccines and biotherapeutics) that have a positive influence on human health. Thus, bacteria are both targets for and producers of biopharmaceuticals. BACELL-HEALTH (2007) is undertaking an integrated and in-depth study of the response of Gram-positive bacteria to stresses encountered by these pathogens during infection and from industrial strains during bioprocesses. The ability of bacteria to detect, respond and resist environmental insult is a key element in their survival and productivity. The physical and chemical insults to which bacteria may be subjected range from generic stresses, encountered by most unicellular organisms in the environment, to specific stresses encountered by pathogens during infection or by engineered strains during industrial bioprocesses. The main objective is to develop a profound understanding of the integrative cell management and associated stress-resistance processes that are essential for sustaining bacteria as effective pathogens or as efficient producers of pharmaceutically active proteins. Studies on relevant generic stress responses are carried out on the genetically highly amenable model bacterium *B. subtilis*, but will be extended to pathogens in the same genus (e.g. *B. anthracis*, *B. cereus*) and related genera (e.g. *L. monocytogenes*, *S. aureus*, *S. pneumoniae*). The aim is to model the regulatory networks that comprise the cell stress management system, identifying key targets for the development of novel anti-infectives and improving the productivity of *Bacillus* for the production of biopharmaceuticals.

Streptomyces lividans

STREPTOMICS (2007) uses members of *Streptomyces* as a protein production platform to study systems biology strategies and metabolome engineering for the enhanced production of recombinant proteins. *Streptomyces* is the largest genus of the order *Actinomycetales*, a group of Gram-positive and generally high GC-content bacteria. Streptomycetes are found predominantly in soil and in decaying vegetation, and most produce spores. Streptomycetes are noted for their distinct "earthy" odour which results from production of a volatile metabolite, geosmin. Streptomycetes are morphologically highly differentiated: they form a coherent mycelium of branching hyphal filaments which under growth-limiting conditions are converted into spores. The major interest in the genus *Streptomyces* is the diversity of secondary metabolites produced by its members, which makes them industrially very important. Many of these secondary metabolites have antibacterial, antifungal and antiviral properties as well as antitumour activity. Several antibiotics produced by streptomycetes are in clinical use (neomycin, chloramphenicol). The first medical use of a *Streptomyces* antibiotic was the treatment of tuberculosis by streptomycin, which took its name directly from *Streptomyces*. Streptomycetes are also used in industry for production of enzymes and recombinant proteins. Streptomycetes are infrequent pathogens, though infections in human such as mycetoma can be caused by *S. somaliensis* and *S. sudanensis* and in plants such as scabies can be caused by *S. scabies, S. acidiscabies*, and *S. caviscabies*.

Protein Secretion in Streptomyces

The biotechnology industry is constantly searching for better hosts for the production of biopharmaceuticals and enzymes of diverse origin. *Streptomyces* has already proved an invaluable host for this purpose, since it can secrete several heterologous proteins in satisfactory amounts. However, in order to optimise strain selection, knowledge is required concerning the following points:

- How protein secretion processes are integrated within the metabolome, and how they interact
- How heterologous protein secretion stresses the metabolome and induces negative cellular cascades

Systems biology provides the means to address these questions by combining biochemical information with genetic and molecular data, leading to a better understanding of the protein secretion mechanism at the cellular level. SREPTOMICS (2007) aims to enhance the production of heterologous proteins, using members of *Streptomyces* as a host. More specifically the project intends:

- To evaluate *S. lividans* as a cell factory for the production of heterologous proteins of interest
- To investigate the transcriptome and proteome of the host strain under different growth conditions, with different expression/secretion vectors, and using different fermentation strategies, in order to identify the genes important for optimal cell performance, with respect to heterologous protein secretion
- To analyse metabolic flux control and flux balance with a view to engineering metabolic pathways found in a *Streptomyces* background, and hence to exploit cellular pathways which provide improved energy transduction, balanced growth and supramolecular assembly
- To engineer better production/secretion strains of *Streptomyces* based on the above, and based on information about secretion bottlenecks that will be identified through the production of mutants, via deletion of a set of selected genes, and muteins, either via direct mutation of specific amino acids, or by directed evolution
- To optimise the protein production process

STREPTOMICS (2007) also plans to develop systems biology strategies and metabolome engineering for the enhanced production of recombinant proteins in members of *Streptomyces*.

Cardiovascular–Pulmonary Diseases and Diabetes

Congestive Heart Failure, Chronic Obstructive Pulmonary Disease and Type-2 Diabetes

BioBridge (2007) will focus on the application of simulation techniques on top of multilevel data, in order to create models for understanding how molecular mechanisms are dynamically related in complex chronic disorders. The hypothesis is that simulation tools may be useful to identify underlying mechanisms of chronic disease phenotypes with systemic effects that are associated with poor prognosis. The project will explore and identify gaps in information, and develop and apply standards for the transfer and filtering of data from existing molecular biology databases and new high-throughput experiments (microarray, in vivo metabolic profiling and proteomics data) into metabolic models of complex diseases. The objectives are twofold: developing software for integrated genomic, proteomic, metabolomic and kinetic data analysis, in order to build a bridge between basic science and clinical practice; understanding the distortion of cellular metabolism that is associated with certain target diseases.

The diseases in question are congestive heart failure (CHF), chronic obstructive pulmonary disease (COPD) and also type-2 diabetes. The available facts strongly indicate that these diseases comprise a cluster of chronic conditions, all of which are associated with nitroso-redox imbalance. The integration of data into a dynamic framework will enable the development of the first kinetic model of the metabolism shared by COPD, CHF and also type-2 diabetes, thereby revealing the common and individual traits of these three complex diseases. Existing computational models have already proved powerful in this context. For example, one of the BioBridge partners has recently developed a statistical framework for analysis of multivariate models from large-scale datasets. This software environment, GALGO (2007), uses a genetic algorithm search procedure, coupled with statistical modelling methods, for supervised classification and regression. GALGO (2007) is relatively easy to use, can manage parallel searches and has a toolset for the analysis of models. Other methods have been developed for biologically driven variable selection, also using a genetic algorithm search strategy. The research programme has the following goals:

- Creation of a structured database for the collection of clinical information relating to COPD, CHF and type-2 diabetes.
- Identification of the metabolic pathways implicated in the target diseases.
- Recording of genomic, proteomic, metabolomic and kinetic information onto the relevant structured databases.
- Development of a software product designed for specific disease-related data searching.
- Development of standards for the different levels of data, which will be useful for their integration from genomic and metabolomic databases, and from specific proteomics and metabolomics profiling experiments, including microarray analysis and stable isotope tracer data. These will be mainly Bayesian networks and multivariate analysis tools.
- Development of protocols for transferring data from the structured databases into dynamic models.
- Using a differential equation approach, the design and development of an innovative simulation environment that will accommodate the dynamic behaviour of complex networks, and in particular the metabolic pathways that are altered by the target diseases.
- Development of generic tools that will be clinically useful beyond the target diseases addressed during the lifetime of the project. BioBridge will also focus on interfacing with end users, in particular clinical researchers and clinicians.

Diabetes Screening

Bioinformatics methods for diagnostic screening are a bottleneck in current biomedical research. While exploratory methods such as statistical hypotheses testing, clustering of gene expression profiles and classification methods have been successful in the detection of molecular markers for interesting diseases, these techniques fail to validate these markers in their gene regulatory context and to integrate other data sources relevant for diagnostic purposes. For these tasks, novel modelling techniques, network analyses and data integration methods are indispensable. The analysis of processes involved in the course of complex polygenic diseases, such as obesity and type-2 diabetes, is in fact a multistep procedure that has to cope with data from diverse experimental functional genomics platforms (gene and protein expression), physiological data, environmental factors and others.

Proteomics and Modelling Approaches

The project SysProt (2007) aims to develop a new paradigm for the integration of proteomics data into systems biology. The goal is to gain relevant knowledge on the biological processes that are important for human health and to use this knowledge for the purpose of disease modelling. In order to achieve this objective, an innovative, explorative biological systems approach (on both the molecular and the physiological level) will be adopted, with a strong focus on protein function and modification. SysProt (2007) will produce proteomics data, indispensable for the identification of novel circulating protein factors, and post-translational protein modifications that are important for the onset, dynamics and progression of complex diseases. Data generation will be complemented by the development of computational analysis methods for these novel data types and the creation of adequate modelling technology. Established mouse disease models will be used with existing benchmarking modules for computational analysis, and the functional genomics platforms will be developed by and accessible to the partners. Newly developed technologies will be demonstrated in a proof-of-principle study within an obesity-induced type-2 diabetes mouse model. An important feature is the integration of phenotypic and physiological parameters with proteomics data and expression profiles from time-course series representing the onset and progression of insulin resistance of type-2 diabetes. Ultimately, the platform will enable medical researchers to combine heterogeneous biomolecular data with physiological and clinically relevant parameters to predict individual predispositions to obesity-induced type-2-diabetes. The main result of the project will be an exploitable prototype that allows medical researchers to make predictions on disease-relevant pathways. The objectives are:

- Model the knowledge about biological objects (genes, proteins and protein complexes) in the context of nutrition and type-2 diabetes in equivalent computer objects
- Integrate heterogeneous data types from proteomics and functional genomics approaches
- Develop and use a prototype framework for the automatic detection and localisation of protein modifications on high-accuracy mass spectrometry data
- Generate specific proteomics and functional genomics data providing the necessary information for disease model generation with an appropriate animal model

- Gain new knowledge on the pathways and marker genes relevant for obesity-induced type-2 diabetes disease progression that will lead to the discovery of novel diagnostic biomarkers for disease susceptibility
- Stimulate perturbations of the disease-relevant pathways
- Develop tools and methods for the correlation of phenotype and genotype
- Accelerate the identification and positional cloning of disease candidate genes by combining gene expression, proteomics, genotype and clinical data
- Set up a knowledge base that integrates all available data and methods as an exploitable product for disease modelling

Neurological Diseases

Systems Biology of the Neuron

The SYMBIONIC (2007) project concentrated on cell and molecular neurobiology and neurophysiology, as related to functional genomics, proteomics, bioinformatics, biophysics and computational biology. The project provided a general assessment of the existing data and know-how in several relevant scientific domains, from neurophysiology to computer science. A good summary is provided by papers presented at the SYMBIONIC workshop on neurogenomics (Symbionic-Workshop 2005), in the following areas:

- Molecular and cellular complexity of the brain.
- The transcriptional landscape of the mammalian genome.
- Genomic organisation of regulatory elements in higher eukaryotes.
- How much (or little!) do we understand about genomes?
- Gene expression in brain development.
- Gene networks governing CNS development.
- Gene expression profiles of neurons.
- Identification of signalling pathways in the developing mammalian brain using genomic approaches.
- Making sense of scents: genomics of vertebrate pheromone receptors.
- The genetics of age-related macular degeneration – recent progress in understanding disease origin.
- Proteomics in neurodegeneration: Huntington's disease.
- The emerging complexity of nerve growth factor processing and Alzheimer's neurodegeneration.

Macular Degeneration

The project EVI-GENEROT (2007) studies diseases that affect the retina such as inherited retinal degenerations and age-related macular degeneration. The goals are to build on understanding of the fundamental molecular and cellular biology of the retina, of its development and the way it is perturbed by genetic mutation, environmental factors and age, so as to:

- Gather and integrate the information on gene function brought about by the numerous human, animal and in vitro models of retinal development and degeneration available.
- Standardise and analyse this information (databases and expression studies)
- Validate the information (functional assays and models)
- Facilitate the design of genome-based therapy that would obviously potentially benefit patients but also validate the pathways and targets identified using the above-described approaches

The goal is to integrate a broad and in-depth understanding of the function and interactions of major cells and genes networks, thereby proposing functional models. The key questions and issues faced in addressing these questions are:

- Obtaining the information provided by the clinical conditions and animal models
- Analysing the information: functional genomic tools
- Validating the information
- Designing genome-based therapy

Modelling Molecular Interaction Networks

VALAPODYN (2007) seeks to further the development of multidisciplinary functional genomics relating to complex biological processes and cellular networks. Mathematical modelling is generally based on the understanding or theory of the way the modelled system behaves, and experimental data of elements of the system and how it reacts under certain conditions. Molecular interaction networks (MINs) are very robust; so it is therefore possible to model them, although this means that a vast amount of data (thousands of genes and proteins) needs to be considered, with highly redundant interactions. Moreover, the different networks behave with non-linear and non-additive responses. All these characteristics therefore necessitate the development of large-scale MIN modelling methods to address the physiopathological processes of many diseases. VALAPODYN (2007) incorporates fundamental genomics research, which will integrate statistical data analysis with real biological information to functionally annotate genes and proteins. Specialised genomics and proteomics databases for MIN building will be used alongside leading microarray and proteomics platform systems, in order to investigate protein–protein interactions and regulation networks; these will help in the identification and validation of important molecules which need to be considered in the MIN. New tools in systems biology will be designed, with the adoption of a dynamic approach to MIN modelling. In the current environment of biological knowledge and techniques, it is not reasonable to expect that simulations performed with such models will accurately reproduce or imitate the real network. However, such models can be used for classifying the effects of physiological/therapeutic actions on the signalling networks in order to generate new hypotheses and help decision making. Such a modelling strategy should lead to the selection of efficient drug targets. However, it will be necessary to validate these targets and therefore the dynamic models, by using RNA interference in cell cultures and animal models. The overall aim is to develop an innovative systems biology approach, in order to model the dynamics of MINs related to cell death and survival in the organism. The specific tasks are:

- Pathways analysis: to analyse functional annotation of genes and proteins, investigation of structure and dynamics of signal transduction and transcription regulatory networks
- Predictive bioinformatics platform for dynamic modelling: to use innovative biomathematics and bioinformatics to integrate experimental MIN data with biological tissue and pathological states data obtained using transcriptomic and proteomic approaches
- Bioinformatics: to set up a highly specialised database on the genomics and proteomics of MIN modelling
- Pathological tissue and animal models: to use novel animal models to evaluate local expression genes/proteins in neurodegeneration
- Microarrays: to set up a network of microarrays, using the Codelink platform
- Proteomics: to form a network of advanced quantitative methods in proteomics technologies, e.g., matrix-assisted laser desorption/ionisation (MALDI), ICAT, two-dimensional polyacrylamide gel electrophoresis (PAGE), heavy peptides isotopic dilution)
- Neuroprotective molecules: to characterise molecules (or combinations of molecules) in the MIN of neurodegeneration, which will have an effect in preventing or curing neurodegeneration

The project could be followed by innovative dynamic modelling of pathological states such as epilepsy and cancer, in order to extend the model applications.

Proteomics Support

The project will also construct a network of advanced quantitative methods in proteomics technologies, including mass spectrometry (electrospray–MALDI) coupled with ICAT labelling, heavy peptides isotopic dilution and two-dimensional PAGE (including two-dimensional difference gel electrophoresis). In addition, the project will set up a network of microarrays (mainly whole genome oligonucleotide gene expression arrays), by applying the new Codelink (Diez et al. 2007) platform. Equally, the novel dynamic models proposed will be validated with innovative pathology models (relating to both animal and cell models). These animal models will be used to evaluate local expression genes/proteins in epilepsy.

Immunology

Immunology Grid Models

ImmunoGrid (2007) is an implementation of a virtual human immune system using grid technologies. It is aimed at simulating immune processes and providing tools for applications in clinical immunology and the design of vaccines and immunotherapies. The developed set of tools will be validated with experimental data and used in clinical applications for development of immunotherapies in cancer and chronic infections. Computational models are required because the immune system is complex and has a combinatorial nature, experimental approaches are expensive, and there are restrictions on the experimentation that can be performed in humans. The target user groups are clinicians and developers of the vaccines and immunotherapies. ImmunoGrid (2007) will provide these users with tools for identification of optimal immunisation protocols. The unique aspect is that it aims at connecting molecule-level interactions (which regulate immune responses) with system-level models (which study the behaviour of the immune system as a whole). Applications include modelling of the natural-size complex system on a large scale and the implementation of individual immune system models across the grid nodes.

Drug Development

Biosimulation in Drug Development

BIOSIM (2007) is a major project that is a pilot for the even more ambitious Innovative Medicines Initiative (IMI 2007). The methods that are currently applied in the development of new medicines suffer from the lack of effective means to evaluate, combine and accumulate biological knowledge. Essential improvements can involve the use of computational models that can provide a dynamic and more quantitative description of the relevant biological, pathological and pharmacokinetic processes (Bertau et al. 2007). Recent BIOSIM (2007) results were summarised in a meeting report (Krishna et al. 2007), which focused on emerging aspects related to the quantitative understanding of underlying pathways in drug discovery and clinical development, i.e. moving from an empirical to a model-based, quantitative drug development process. The BIOSIM (2007) research activities include the following.

Mass-Spectrometric Resolution of Protein Phosphorylation in Hormonal Signalling

The reversible phosphorylation of specific proteins participates in the regulation of virtually all aspects of cell physiological processes and development. The importance of this process is illustrated by the many hundreds of protein kinases and phosphatases detected in eukaryotic genomes. Reversible protein phosphorylation is the major mechanism for external control. The project is working to:

- Characterise adipocyte phosphorylation and changes in the phosphorylation process in response to insulin treatment
- Identify proteins that undergo reversal of phosphorylation and their phosphorylation sites
- Determine the stoichiometry of hormone-induced protein phosphorylation and the spatio-temporal variations in this process
- Examine cells from patients exhibiting insulin resistance to compare their phosphorylation processes with those of normal adipocytes
- Develop mathematical models that can integrate and evaluate the large amount of data generated by the experimental activities

Model development will occur along several lines:

- Development of algorithms and software for automatic phosphopeptide identification and calculation of phosphorylation stoichiometries from mass spectrometry raw data
- Development of tools to visualise the multidimensional arrays of phosphorylation states of hundreds of proteins that change in time and space as well as within the individual protein
- Development of nonlinear dynamic analyses to help us understand the mechanisms behind the changes in phosphorylation patterns

The dynamic spatio-temporal mapping and visualisation of the information flow in response to insulin can provide a hitherto unknown level of understanding of how this hormone works, in health and disease.

Metabolic Fates of Pharmaceuticals in Living Cells

Pharmaceuticals are often polyfunctional molecules and their metabolic fates are difficult to predict. In the quest to replace animal tests by other approaches, microbial models of drug metabolism have been applied for more than 30 years. However, owing to the complexity of the molecules, the results obtained have not yet allowed us to extract general principles of eukaryotic degradation of drugs. Hence, there is a significant demand for a modelling system that can describe the most basic interactions of the enzymes involved in drug biotransformation. The PharmBiosim approach aims at identifying general principles of drug metabolism – also for complex pharmaceuticals – by attributing enzyme actions and xenobiotic stress phenomena to functional groups and, thereby, reducing the complexity of the drug molecule to those functionalities provoking chemical modifications. The programme for this project includes:

- Initial investigations of the metabolic fates of small monosubstituted and trisubstituted organic compounds. This allows direct attribution of cellular responses to the presence of functional groups in an organic substrate.
- Attempts to formulate general principles for xenobiotic metabolism.
- Establishment of a model for the bioinformatics of ethyl acetoacetate.
- Studies of the biotransformations of ethyl 2-chloroacetoacetate, ethyl 4-chloroacetoacetate and 4,4,4-trifluoroacetoacetate.

Microcompartments Associated with Microtubular Networks

Glucose is the principal energy source in brain and red blood cells and it is metabolised via glycolysis. In the brain the microtubular system is of special importance and it comprises almost all neuronal volume and surface area. Glucose conversion is characterised by multiple pathways and the consequence of microcompartmentation of enzymes by tubulin in brain extract. The micropathway analysis is extended to the whole glycolytic and pentose phosphate pathways, accounting for the form-specific (isoform, oligomeric form and conformation) enzyme associations and the effects of other endogenous and exogenous factors (e.g. drugs) present in the brain tissue. This programme involves the:

- Extension of the micropathway analyses to the full glycolytic and pentose phosphate pathways and investigation of the consequences of microcompartmentation of enzymes by tubulin in brain cells
- Development of a mathematical model that includes the mutant enzyme functionalities and that can be used to examine whether the mutation alone can account for the altered glycolytic flux observed in the presence of tubulin
- Structural analyses of the TPPP/p25 protein using both isolated protein from bovine brain and recombinant human protein
- Investigations of the physiological and pathological properties of the new protein family TPPP/.

Regulation of Pancreatic α- and β-Cells

The pancreatic cells play a central role in blood glucose homeostasis. Glucose increases insulin secretion from β-cells and somatostatin release from δ-cells, but suppresses glucagon release from α-cells. In all three cell types, exocytosis is stimulated by Ca^{2+} influx and the local elevation of Ca^{2+} concentration that results from changes in the electrical activity. Thus, electrical activity plays an essential role in the regulation of islet hormone release. Indeed, it is because of their different complements of ion channels that glucose has opposite effects on β- and α-cell secretion. Research areas include:

- Complete characterisation of ion channels expressed in pancreatic islet cells (molecular and biophysical characterisation) in situ
- Regulation of the islet cell ion channels by cytoplasmic metabolites and intracellular Ca^{2+} concentrations
- Relationship between electrical activity and changes in the cytoplasmic free Ca^{2+} concentration.
- Control of exocytosis by rapid and extensive increases in the near-membrane Ca^{2+} concentration
- Replenishment of the release-competent pool of granules for release
- Modelling the kinetics of peptide release via the narrow pore (fusion pore) connecting the granule lumen to the extracellular space
- Elucidation of the role of paracrine (hormonal) and electrical (gap junctions) signalling on electrical activity and secretion
- Establishment of the mechanisms by which small changes in the metabolic regulation and/or activation/inactivation properties of islet cell ion channels (such as those occurring in diabetes) influence islet cell electrical activity and secretion
- Modelling of the effects of drugs through drug-mediated modifications of the electrical activity

The individual islet cells are modelled in several stages:

- In the case of the β-cell, for example, first construct a mathematical model of electrical activity based on experimentally measured ion channel parameters.
- Subsequently, incorporate a model of β-cell metabolism, to address how electrical activity is modulated in response to glucose and other nutrients and, conversely, how electrical activity influence metabolism.
- The next stage will be to generate similar models for the other islet cell types.
- Finally, the separate models of the individual islet cells will be synthesised to produce a unified model of the whole islet.

Neuronal and Systemic Models of Mental Diseases and Sleep Regulation

Mental disorders are among the most common diseases. Depressive disorders have a prevalence of about 15%. Besides substantial individual suffering this also entails enormous socioeconomic costs. Depressive disorders can be treated to some extent by antidepressant drugs, an often-used type being the selective serotonin reuptake inhibitors. These substances enhance the neuronal actions of serotonin.

Serotonin is a neurotransmitter which contributes to information processing in many areas of the brain, including those that are involved in mood control and sleep–wake cycles. Remarkably, the antidepressant drug effects occur with long time delays (2 weeks or more), which indicates that the desired effects are caused by secondary changes. To define the functional principles which are relevant for antidepressant drug treatment, computer models are developed at different functional levels and in connection with a number of different phenomena. Optimisation of the effective, safe and selective antidepressant drug treatment is a major issue which can benefit from the use of computational approaches including data banking, biosimulation and system analysis. The approach addresses two specific issues – regulation and plasticity – with respect to different biological levels as well as with respect to different levels of analysis (i.e. detailed vs. simplified simulations):

1. Network regulation: One of the challenges arising from the different pathophysiological factors, besides characterisation of the major pathobiological players, is to understand the regulatory issues and network dynamics that arise from the interplay of these factors.
2. Adaptation and plasticity: The second specific task, closely associated to regulatory mechanisms, is understanding of the long-term adaptive and maladaptive effects relevant for the pathological processes of mood disorders as well as with regard to the actions of drugs.

Synchronisation of Nephron Pressure and Flow Regulation

The kidneys play an important role in regulating the blood pressure and maintaining a proper environment for the cells of the body. The process of glomerular ultrafiltration is highly sensitive to variations in the blood pressure, and a proper regulation of the excretion of water and salts involves mechanisms that can compensate for variations in the arterial blood pressure. Experimental investigations have shown that these interactions can cause neighbouring nephrons to operate in synchrony. A detailed model has been developed of two coupled nephrons. This model can account for all the observed synchronisation phenomena. The questions to be addressed are:

- What is the length scale of the synchronisation phenomena, i.e., to what extent can the synchronisation spread to nephrons situated along adjacent interlobular arteries or to juxtamedullary nephrons?
- What role do synchronisation phenomena play in connection with development of hypertension? It is well known that the vascularly mediated interactions are stronger for hypertensive rats than for normotensive rats.
- How variable are the synchronisation patterns, and to what degree are they affected by antihypertension drugs or gap junction inhibitory peptides?

The analysis involves:

- Development of new experimental techniques to investigate the synchronisation patterns of surface nephrons, and techniques to study the regulatory dynamics of juxtamedullary nephrons.
- Development of new data analysis techniques to reveal the details of various synchronisation states.
- Formulation of a large-scale simulation model of 20–30 interacting nephrons, a model that can account both for the complex dynamics of the individual nephron and for the complex structure of the arteriolar tree.
- Detailed modelling of signal propagation along the arteriolar vessels. This part of the project will be performed in collaboration with a project on coupled smooth muscle cells.
- A series of experiments to investigate the influence of various antihypertension drugs or gap junction inhibitory peptides on both the function of the individual nephron and the coupling between the nephrons.

The coupled nephron model will provide an important contribution to understanding:

- Kidney regulation
- How an ensemble of biological oscillators, each displaying complicated nonlinear dynamic phenomena, can interact to produce different forms of coherent behaviour on a higher structural level
- The possible effects of antihypertension drugs or gap junction inhibitory peptides on kidney function

Models of Full-Scale Cardiac Arrhythmias

A platform has been developed where it is possible to construct models of full-scale cardiac arrhythmia, including re-entrant arrhythmias using detailed biophysical cell models. To date most work in this area has been done with greatly simplified cell models. The computing resources now available make it possible to attempt such reconstructions with more biophysically detailed models. The following arrhythmia mechanisms are ready for incorporation into tissue and whole organ models:

- Sodium channel mutations leading to early after-depolarisations (EADs).
- Drug-induced EAD class III drugs, acting to inhibit potassium channels, are well known to induce torsade de pointes arrhythmias.
- Sympathetic overdrive, including exercise.
- Delayed after-depolarisations.
- Slowed conduction.

The role of anatomical detail, including pathological changes, and of electrophysiological inhomogeneity need to be assessed in all these forms of arrhythmic mechanism.

Spatio-temporal Organisation of Intracellular and Intercellular Ca^{2+} Dynamics

Calcium is a widespread second messenger, mediating important physiological responses in all types of organisms, from bacteria to specialised neurons. It has been known for about 15 years that the calcium increases induced by an external stimulation are highly organised, both in time and in space. Indeed, the rise in cytosolic calcium concentration occurs in the form of repetitive calcium spikes. These calcium oscillations are observed in most cell types and are considered as a prototype of an oscillating system in cellular biology. Moreover, each calcium spike is also organised at the spatial level; the rise in calcium concentration is first restricted to a portion of the cell, and later invades the whole cell as a wave. It is well established that calcium oscillations result from a periodic exchange of calcium between the cytosol and the internal calcium stores (endoplasmic reticulum). In response to the external stimulation, inositol 1,4,5-trisphosphate (InsP3) is synthesised in the cytoplasm. InsP3 receptors are calcium channels located in the membrane of the endoplasmic reticulum. Periodic release of calcium from the reticulum can be ascribed to the autocatalytic regulation by which calcium can activate its own release through the InsP3 receptors. A theoretical model has been developed based on the coupling between several oscillators (i.e. the individual cells of the multiplet), whose dynamics is described by a model previously proposed to account for intracellular calcium oscillations. Numerical integration of the model shows that it is possible to coordinate calcium spiking among connected hepatocytes when it is assumed that InsP3 can somewhat diffuse through gap junctions; calcium spiking, however, occurs with a slight phase shift among connected cells, giving rise to the appearance of a phenomenon of wave propagation. The direction of the wave is imposed by the direction of the sensitivity gradient. Models of intracellular and intercellular calcium waves will also be developed for pancreatic cells and for smooth muscle cells in the arteriolar wall.

Modelling of Molecular Regulatory Mechanisms of Circadian Rhythms

Circadian rhythms occur with a period close to 24 h in all eukaryotic and some prokaryotic organisms. These rhythms have a profound influence on human biological processes. Experimentally based theoretical models have been developed for circadian rhythms. New aspects of these rhythms will be investigated by focusing on the detailed molecular mechanism of circadian clocks in *Drosophila* and mammals, with extension to the origin and consequences of circadian disorders in human physiology. The model proposed describes in detail the molecular mechanism responsible for circadian rhythms in *Drosophila*. This model incorporates the effect of light that induces degradation of the TIM protein. By means of this model one can therefore determine the effect of a light pulse. Critical light pulses can suppress circadian rhythms in a permanent manner and the rhythm can be restored by a second perturbation identical to the first pulse. The model proves useful in allowing quantification of the duration and amplitude of the effect of a light pulse necessary to suppress rhythmic behaviour. Remarkable progress has also recently been made in unravelling the molecular mechanism of the circadian clock in mammals, where the circadian pacemaker is located in neurons of the suprachiasmatic nuclei in the hypothalamus.

An important aspect of circadian rhythms pertains to the implications of these rhythms for pharmacology. Given that most physiological functions vary in a circadian manner, it is not surprising that the toxicity of many drugs as well as their efficacy vary in the course of the day with a circadian period. This aspect has long remained practically unnoticed in pharmacology but is increasingly gaining interest as shown by the slow but sure development of chronopharmacology, whose goal is to determine the optimal timing for administration of medications, as a function of the physiological rhythms of the patient. The most convincing advances in this field are probably those made in cancer therapy, where multicentric trials of phase III are under way for the treatment of colon cancer. Modelling studies based on the pharmacokinetics of the drug and of the circadian rhythms involved in drug action and degradation should contribute to optimising the patterns of drug delivery.

Deep Brain Stimulation and Medication

In several neurological diseases like Parkinson's disease or essential tremor brain function is severely impaired by synchronisation processes. Parkinsonian resting tremor appears to be caused by a population of neurons located in the thalamus and the basal ganglia. These neurons fire in a synchronised and intrinsically periodic manner at a frequency similar to that of the tremor, regardless of any feedback signals. In contrast, under physiological conditions these neurons fire incoherently. In patients with Parkinson's disease this cluster acts like a pacemaker and activates premotor areas and the motor cortex, where the latter synchronise their oscillatory activity. In patients with advanced Parkinson's disease or essential tremor who no longer respond sufficiently to drug therapy, depth electrodes are chronically implanted in target areas like the thalamic ventralis intermedius nucleus or the subthalamic nucleus. Electrical deep brain stimulation (DBS) is performed by administering a permanent high-frequency (above 100 Hz) periodic pulse train via the depth electrodes. High-frequency DBS has been developed empirically, and its mechanism is not yet fully understood.

The goal is to study the interplay between medication (e.g. DOPA) and demand-controlled DBS in order to improve the clinical outcome. A modelling and computer simulation approach is used to investigate the impact of variations of the drug concentration on both the stimulation outcome and how the stimulator may compensate for this by appropriate learning algorithms:

- Derive a physiologically realistic model of a neural network in the typical target areas of DBS (e.g. subthalamic nucleus and thalamus).
- Investigate how electrical stimulation via macroelectrodes can be modelled appropriately on such a microscopic level of description.
- Study how variations of the blood concentration of drugs (such as DOPA) show up in terms of varying model parameters (e.g. synaptic strength).
- Incorporate learning algorithms into the demand-controlled stimulation techniques in order to compensate for physiologically realistic variations of model parameters. In particular, test the performance of different demand-controlled stimulation techniques (see above) under the influence of such variations and supported by appropriate learning algorithms.

Biological Networks, Data Analysis and Pharmacokinetic Models

A central theme is the integration of genomes and high-throughput data with mathematical modelling of cellular processes. Statistical dependences and periodicities are studied in complete genomes, reproducibility of DNA chips and large two-dimensional gels are quantified, and tools are developed for the analysis of promoters of co-regulated genes. These bioinformatics techniques are applied to specific cellular systems as a prerequisite of mathematical modelling. The following themes are investigated:

- Regulatory networks: statistical links between models and data
- Modelling transport and protein sorting across cell membranes and compartments
- Virtual populations and drug development
- Non-linear signal analysis

Modelling Human Metabolism, Body Weight Regulation and the Treatment of Diabetes

Modelling human metabolism and body weight regulation are key elements in approaching the treatment of diabetes. Detailed models allow the simulation of various drug approaches.

Live Cell Imaging by Use of Interference Microscopy

This research directly studies the effect of different diseases on the function of red blood cells, immune cells, nerve cells, etc. The technique also allows the examination of the influence of various drugs on the cellular processes, and several simultaneous intercellular processes and their interactions to be followed without disturbing the cells.

Application of Methods from Non-linear Dynamics to Describe Complex Cellular Phenomena

Non-linear dynamics are involved in studying chronotherapy, DBS, analyses of depression, modelling of cellular interactions, etc. Mastering of this area of modern mathematics/physics is a distinctive characteristic of the BioSim (2007) network.

Implementation in the Seventh Framework Programme

Innovative Medicines Initiative

IMI (2007) , is a major joint effort planned by the European Commission and the European Pharmaceutical Industry Association (EFPIA 2007). The objective of IMI is to support the faster discovery and development of better medicines for patients and to enhance Europe's competitiveness by ensuring that its biopharmaceutical sector remains a dynamic high-technology sector. IMI has been accepted as a Joint Technology Initiative by the European Council and European Parliament in 2008. Extremely large contributions are expected to be made by both the European Commission and the European pharmaceutical industry. The research projects will not develop new drugs per se but will generate new knowledge about diseases and new tools and technologies, thus better underpinning, improving and accelerating development of new therapies. The IMI (2007) Strategic Research Agenda goals include:

- Predicting safety
- Predicting efficacy
- Bridging gaps in knowledge management
- Bridging gaps in education and training

These goals will be updated as necessitated by scientific advances. In the publications and research agenda, there is a section about plans for knowledge management.

The Innomed (2007) research project in FP6 (2007) can be considered as a prototype project for IMI (2007). Its PredTox activity aims at studying toxicological aspects of new treatments at an early phase. This involves the construction and delivery of an integrated database populated with data from in vivo experiments of compounds with known toxicity profile. It will include traditional end points supplemented with information from newer techniques, i.e. transcriptomics, metabolomics and proteomics. This will among other things demonstrate how the different pharmaceutical companies can share scientific information between themselves, academia and biotechnology companies in order to implement the application of new tools to aid decision making in preclinical safety. The AddNeuroMed activity is using the consortium's expertise in analytical techniques, preclinical and clinical development to provide technologies to facilitate and accelerate the delivery of safe and effective medicines whilst addressing the issues of:

- Absence of diagnostic markers
- Lack of biomarkers of progression
- Lack of biomarkers of response/non-response

Seventh Framework Programme Research Projects in the Systems Biology of Disease

A series of related topics in the first call for proposals FP7-CALL-HEALTH-2007-A (2007) is leading to the following projects:

Immunology: *Modelling of T-cell activation.*
SYBILLA: Systems Viology of T-cell Activation in Health and Disease. See the end of Chap. 3 for a detailed description.
Apoptosis: *Developing an integrated in vitro, in vivo and systems biology modelling approach to understanding apoptosis in the context of health and disease.*
APO-SYS: Apoptosis systems biology applied to cancer and AIDS. An integrated approach of experimental biology, data mining, mathematical modelling, biostatistics, systems engineering and molecular medicine. See the end of Chap. 8 for a detailed description.

Chapter 8
Cancer

Abstract Cancer has a separate chapter devoted to it because of its nature which ties it particularly to bioinformatics and systems biology approaches, and because of its extreme genetic diversity and multistage complexity. Nevertheless, there are many common themes which are explored here. The nature of cancer is reviewed, and then systematically explored, considering in turn the roles of tumour viruses, cellular oncogenes, growth factors and their receptors, cytoplasmic signalling circuitry, cell cycle, tumour suppressor genes, p53 and apoptosis, cell immortalisation, tumourigenesis and senescence, multistep tumourigenesis, genomic integrity and the development of cancer, angiogenesis and lymphangiogenesis, metastasis and tumour immunology and immunotherapy. The implementation of dedicated programmes in the Seventh Framework Programme are discussed.

Introduction

Cancer Research Programmes

Cancer has a separate chapter devoted to it because of its nature which involves progressive DNA sequence mutations and chromosomal alterations and which originates from a single cell, tying it particularly to bioinformatics and systems biology approaches. As well as major national, charitable and international programmes, the European Commission within the Sixth Framework Programme had a separate and major programme devoted to cancer (Manoussaki 2006).

Another major set of collaborative efforts is represented by the Integrative Cancer Biology Programme (ICBP 2007) of the US National Cancer Institute (NCI 2007). In addition to funding a number of interdisciplinary centres, the ICBP (2007) centres interact and collaborate with other NCI programmes and external groups. This chapter concentrates on large-scale European collaborative projects, so the reader is referred to the ICBP (2007) website to appreciate the

F.B. Marcus, *Bioinformatics and Systems Biology*,
DOI 10.1007/978-3-540-78353-4,

programmes at its research centres, each one involving collaborations, including:

- Case Western Reserve University studies DNA repair and its relation to drugs from a clinical perspective.
- Dana Farber Cancer Institute focuses on creating predictive models for cancer defined in terms of cellular modules (such as pathways), specifically on the kinases.
- Duke University is focused on the development of data and computational tools that will substantially advance our understanding of critical cell signalling pathways, primarily on the Rb-E2F pathway with additional interest in the intersection with Ras, Myc and p53.
- E. O. Lawrence Berkeley National Laboratory's goal is to develop experimental and computational strategies to predict individual responses to cancer therapies targeted along the Raf/MEK/ERK signalling pathway.
- The goal of the Center for the Development of a Virtual Tumour (CVIT 2007) of Massachusetts General Hospital is the design and development of a module-based toolkit for cancer research guided by a complex systems approach to integrate multiple levels of information about cancer.
- Massachusetts Institute of Technology research focuses on (1) mitogenic signaling networks, (2) DNA repair and (3) migration signalling networks.
- The Ohio State University investigates how the epigenome interacts with the genome in the genesis and the progression of human cancers.
- Stanford University School of Medicine aims to understand the mechanisms driving the transformation of follicular lymphoma to the more aggressive diffuse large B-cell lymphoma.
- Vanderbilt University Medical Center focuses on the parameterisation of the mathematical models of cancer (i.e. hybrid discrete continuous) at the cellular, multicellular and organ biology scales.

The Nature of Cancer

Cancer is a very complicated multistage, multitissue disease (Cassidy et al. 2002). Key principles governing cancer progression were described by Hanahan and Weinberg (2000), and the work was extended by Hahn and Weinberg (2002). Hanahan and Weinberg (2000) propose "that the vast catalogue of cancer cell genotypes is a manifestation of six essential alterations in cell physiology that collectively dictate malignant growth: self-sufficiency in growth signals, insensitivity to growth-inhibitory (antigrowth) signals, evasion of programmed cell death (apoptosis), limitless replicate potential, sustained angiogenesis, and tissue invasion and metastasis." They also highlight genome instability as an enabling characteristic, including the high variability of pathways leading to cancer, and the multiplicity of cell types within tumours. Extensive advances in understanding have been made at the molecular level (Macdonald et al. 2004), and in applying bioinformatics and systems biology to cancer research (Nagl 2006). Nagl (2006) provides an excellent description of various computational approaches, including the roles of cancer as a system, integrated informatics platforms, mathematical models, computer simulation of tumours, structural bioinformatics, and in vivo modelling, tissue resources and data. Sanga et al. (2006) provide a comprehensive review of mathematical modelling of cancer progression and response to chemotherapy. They demonstrate that simulators of various aspects of cancer need to function at multiple levels from genetics to tumours, often with interactions between the various levels. An excellent and current overview of the biology of cancer was provided by Weinberg (2007). He illustrates the need for an integrated approach to understanding cancer. Key mechanisms involve cell signalling, DNA repair, the cell cycle, apoptosis, gene transcription and splicing, infection and immune response. This chapter organises collaborative research contributions similarly to Weinberg-Contents (2007), and investigates various aspects of this process. His introduction makes the following points:

- Tumours can be either benign (localised, non-invasive) or malignant (invasive, metastatic). The metastases spawned by malignant tumours are responsible for almost all deaths from cancer.
- Virtually all cell types in the body can give rise to cancer, but the most common human cancers are of epithelial origin – the carcinomas.
- Cancers seem to develop progressively, with tumours demonstrating different gradations of abnormality along the way from benign to metastatic.

Weinberg (2007) makes several other central observations: Since metazoa only arose once about 700 million years ago, many signalling mechanisms developed then have been conserved. Therefore, findings from model organisms about some of these mechanisms may be applied to humans. Moreover, even though there are hundreds and perhaps thousands of varieties of cancer, there are some universal factors that serve as central research themes, in particular the pRb and p53 proteins, which are products of tumour-suppressor genes that are of pre-eminent importance in human tumour pathogenesis, since the signalling pathways they control are deregulated in the great majority of cancers. Despite the extreme complexity of the processes involved, there is optimism that we may be able to formulate and quantify some organising principles that place all types of human tumours under a common conceptual roof, extending from the six general types of Hanahan and Weinberg (2000). At some point, we may understand in detail how each regulatory circuit operates to control cell phenotype and how to model the operations of each mathematically.

Challenges of Cancer Research

Cancer is, after decades of research, still a devastating disease, responsible for roughly one quarter of deaths. Cancer is clearly one of the most urgent health research problems, and therefore deserves a high priority owing to the large number of deaths, the enormous human suffering, and also related health care and other associated costs. While progress has been made in the treatment of rare childhood cancers, less progress has been made than anticipated in the treatment of the common forms of cancer, responsible for most of the death toll. Even new anticancer drugs like Herceptin or Glivec are successful for only a fraction of patients. Essentially, the two main causes of cancer are genetic predisposition and environmental influence, including infection and inflammation. However, on a more analytical and molecular level, the ontogeny (origin and development) of cancer is less evident, and both clinical as well as basic research suggest that cancer is the result of an accumulation of many factors that promote tumour growth and metastasis. Because of this complexity of cancer, a more systematic approach is needed for understanding and improving further cancer treatment.

Relevance of Collaborative Research Projects

The projects discussed in this chapter have activities which contain bioinformatics and systems biology approaches, and are highly relevant to or directly dedicated to cancer research. Grouping together these efforts demonstrates a comprehensive programme in the systems biology of cancer. Some of the projects such as ESBIC-D (2007) are preparing for major linkages between researchers. Others such as COMBIO (2007), which addresses the dynamic behaviour of the p53 system, are able to gain insights into particular systems by an integrated approach that is not necessarily accessible to the work on individual aspects of the system.

Nature of Cancer and Biology and Genetics of Cells and Organisms

Systems Biology of Cancer

ESBIC-D (2007) undertakes activities to combat multigenic complex diseases, in particular breast cancer. The goal is to establish a framework for a systems biology approach to cancer. This framework consists of different data generated from clinical phenotypes as well as tools that are able to analyse and integrate these data to perform network-based test studies on several aspects of cancer systems biology. The project unites groups with a strong clinical focus, with experience in high-throughput functional genomics, and those with computational and systems biology resources. Special attention is paid to the analysis of discrepancies and coherencies in the data sources. A cancer-relevant model repository is being established consisting of known pathways and gene regulatory networks (such as apoptosis, retinoblastoma and epidermal growth factor receptor pathways) associated with cancer, the role of specific mutations or other changes in key genes/gene products in these pathways, for example PTEN loss-of-function mutation, and, as far as available, detailed clinical data with special emphasis on the influence of different anticancer drugs on these pathways.

Experimental and Clinical Data and Theoretical Models

The strong interaction of clinical and experimental data with theoretical computer modelling can best be achieved in an interdisciplinary and collaborative approach. Important research areas are being identified that combine experimental and clinical data with theoretical models and which will guide further analyses and approaches. Attention is given to *in silico* models of cancer-related (e.g. signalling) pathways which analyse the feedback of theoretical models and experimental data as well as the construction of a complete human metabolic network in order to test responses to drugs and chemical treatments. Key activities include the collection and provision of biomolecular, computational and clinical information for cancer; the collection and implementation of computational models of cancer-relevant processes; the identification of relevant and crucial parameters for future systems biology approaches to cancer; and training activities.

Tumour Viruses

Role of Chronic Infections

INCA (2007) investigates the role of chronic infections and tumour viruses in the development of cancer. About 17% of human cancer cases occurring worldwide are caused by one of six human viruses:

1. Human papilloma virus (HPV)
2. Epstein–Barr virus (EBV)
3. Hepatitis B virus (HBV)
4. Hepatitis C virus (HCV)
5. Human herpes virus 8 (HHV8, Kaposi's sarcoma herpes virus)
6. Human T-cell leukaemia virus I (HTLV-I)

Also the bacterium *Helicobacter pylori* causes many cases of stomach cancer, and further unknown infectious agents that contribute to cancer may exist. INCA aims towards a better understanding of:

- Molecular and cellular mechanisms of cancers caused by these infectious agents
- Mechanisms of long-term persistence of these infectious agents in apparently healthy hosts
- Genetic factors that contribute to cancers associated with infection

On the basis of this knowledge INCA will develop and validate in vivo models to study chronic inflammation and cancer progression and new diagnostic procedures to identify infected people likely to develop cancers. INCA will concentrate on the four themes that are key to all the other cancer-inducing infectious agents on which the research work will be carried out:

1. Persistence
2. Predisposing factors
3. Intracellular mechanisms
4. Prevention and therapy

Bioinformatics and Technology Platform

The partners will combine their most advanced methodological tools to create a technology service platform provided as a central facility for the INCA project. This will have several benefits:

- Availability of a much more powerful technology platform than the individual partners have at their disposal
- Homogenisation of experimental data formats
- Centralisation of data storage and analysis and central evaluation of the potential as well as refinement of this method

The Technology Service Platform will be composed of a microarray gene expression facility, an RNA interference high-throughput facility, a data warehouse, existing microarray data from infection experiments, proteomic technology and novel analysis software tools.

Cellular Oncogenes

Oncogenes Mutation Databases

BioSapiens-WP109 (2007) is establishing methods for the analysis of functional consequences of cancer-associated oncogenes mutations in the context of collaborations with cancer genomics groups. The initial phase of the work will determine the technology to collate and organise functional information on mutations in proteins detected in cancer-screening projects, including tools to deliver the information to experimental biologists interested in these proteins/genes. Collaborations will be established with experimental groups producing high-throughput screening of cancer tissues able to follow up with additional experiments and additional information and insight on the experimental details. Initial considerations point to two possible scenarios: the analysis of new datasets on mutations detected in 580 human kinases in a set of 200 cancer samples (primary tumours mainly) and the analysis of a large set of mutations detected in 4,000 selected genes in samples and cell lines from breast and lung cancer, and melanomas. In these cases, and in other potential collaborations with NCI (2007) funded cancer genome projects, BioSapiens (2007) will organise the analysis of the potential impact of the corresponding mutations in their protein context and in their network context (protein interactions and signalling pathways), with the goal of producing lists of candidate genes – and their mutations – likely to be specifically associated with the various cancer types. These lists will be the entry point for additional high-throughput screening, focusing on selected cancer types, and will also open the door to collaborations with experimental groups interested in those proteins. The direct computational analysis of the individual proteins/ mutations will have to be complemented by a detailed analysis of the biological context (pathways, interactions networks) in the context of the known biology for each biological scenario.

There is also close integration with and involvement of the Cancer Genome Project (2007) of the Wellcome Trust Sanger Institute (WTSI 2007). The identification of genes that are mutated and hence drive oncogenesis has been a central aim of cancer research since the advent of recombinant DNA technology. The Cancer Genome Project is using the human genome sequence and high-throughput mutation detection techniques to identify somatically acquired sequence variants/mutations and hence to identify genes critical in the development of human cancers. The following data resources are available, and much more:

- Cancer Gene Census: Mutated genes causally implicated in human cancer
- COSMIC: Catalogue of somatic mutations in cancer
- Cancer Genome Project Resequencing Studies: Somatic mutations from systematic large scale resequencing of genes in human cancers
- Cancer Genome Project Cancer Cell Line Project: Resequencing of known cancer genes and other analyses of human cancer cell lines
- Cancer Genome Project Copy Number Analysis in Cancer: Analysis of copy number and loss of heterozygosity in cancer cell lines and primary tumours
- Cancer Genome Project Trace and Genotype Archive: Archive of sequence traces and genotype data generated by the group

The publications page of the website contains a very large amount of additional information.

These projects are also carried out as major collaborations with the US collaborative research project The Cancer Genome Atlas (TCGA 2007) of the NCI (2007) and the National Human Genome Research Institute (NHGRI 2007). TCGA (2007) establishes an integrated network of clinical sites, core resources and specialised genome characterisation and genome sequencing centres that work together to form a system that selects genes and regions in order to drive high-throughput cancer genome sequencing. The major organisational and functional components of the pilot project are:

- Biorepositories contributing to TCGA
- TCGA biospecimen selection process
- Request for information
- Human cancer biospecimen core resource
- Cancer genome characterisation centres
- Genome sequencing centres
- Data management, bioinformatics and computational analysis
- Technology development

The informatics component of TCGA involves developing the best ways to collect, store and distribute the clinical and genomic data generated by the project. Among the issues that are being considered are the development of data standards and controlled vocabularies for each new technology, the establishment of an informatics pipeline for data to flow from production centres to a central repository, the creation of portals for basic and clinical researchers to easily access the TCGA data and the encouragement of new computational approaches to analyse the data. TCGA will continue to leverage the resources from Cancer Biomedical Information Grid (CABIG 2007), which has developed the many resources that will be used in the pilot project, such as common data elements, metadata and middleware to enable interactions among distributed databases, and which provides a technical means to support the distribution of data and access to analytical tools for genomic data.

Alternative Transcripts as Cancer Markers

In general, important bioinformatics techniques are being developed to analyse marker genes/transcripts in human cell lines (Tiffin et al. 2005). Genome-wide experimental techniques such as microarray analysis, serial analysis of gene expression, massively parallel signature sequencing, linkage analysis and association studies are used extensively in the search for genes that cause diseases, and often identify many hundreds of candidate disease genes. Selection of the most probable of these candidate disease genes for further empirical analysis is a significant challenge. Additionally, identifying the genes that cause complex diseases is problematic owing to low penetrance of multiple contributing genes. One experimental project in ATD (2007) focuses on the selection of candidate alternative transcripts for human tumours, including colorectal, cervical and lung cancer specific splice patterns and genes that are functionally interesting in cancer settings. Verification of alternative transcript cancer markers was performed by reverse-transcription polymerase chain reaction (RT-PCR) techniques, using cell lines derived from neoplastic human tissues of the colon, the cervix and the lung. ATD (2007) uses a novel bioinformatics approach that selects candidate disease genes according to their alternative transcript expression profiles. It uses the anatomical eVOContology (2007) to mine available human gene expression data for cancer-specific events. To demonstrate that the method is successful and widely applicable, 424 splice events were chosen for RT-PCR experiments; 230 candidate splice events predicted to be cancer-related and 194 corresponding reference events which should be detectable in normal and/or neoplastic cells. The experiments showed a rather cancer-related expression for 73 of the 230 (31%, 73%) candidate splice events comparing normal human tissues and human cancer cell lines (Gautheret, 2007). This approach facilitates direct association between genomic data describing gene expression and information from biomedical texts describing disease phenotype, and successfully prioritises candidate genes according to their expression in disease-affected tissues.

Growth Factors and Their Receptors

Cell Growth Modelling

UNICELLSYS is a new FP7 (2007) project (see Chap. 3) that will use a systems biology approach to cell growth and proliferation as controlled and coordinated by extracellular and intrinsic stimuli. Achieving an understanding of the principles with which biomolecular systems function requires integrating quantitative experimentation with simulations of dynamic mathematical models. UNICELLSYS will study cell growth and proliferation at the levels of cell population, single cell, cellular network, large-scale dynamic systems and functional modules.

AGRON-OMICS (2007) also focuses on growth factors in plant model organisms, see Chap. 4.

Cytoplasmic Signalling Circuitry

Ras/Raf/MEK/ERK and JAK/STAT Signalling

COSBICS (2007) considers two important signalling systems, the Ras/Raf/MEK/ extracellular signal-regulated kinase (ERK) and JAK/STAT pathways. Combining mathematical modelling with biology, the project will improve our understanding of how these are subverted in cancerous tumour cells.

MAPK Signalling

In a discussion of mitogen-activated protein kinase (MAPK) signalling pathways in cancer (Dhillon et al. 2007), COSBICS (2007) has shown that cancer can be perceived as a disease of communication between and within cells. The aberrations are pleiotropic (multiple phenotypic traits), but MAPK pathways feature prominently. Cancerous mutations in MAPK pathways mostly affect Ras and B-Raf in the extracellular signal-regulated kinase pathway. Stress-activated pathways, such as

Jun N-terminal kinase and p38, largely seem to counteract malignant transformation. The balance and integration between these signals may widely vary in different tumours, but are important for the outcome and the sensitivity to drug therapy.

Wnt and ERK Pathways

Kim et al. (2007) noted that the Wnt and the ERK pathways are both involved in the pathogenesis of various kinds of cancers. Recently, the existence of crosstalk between Wnt and ERK pathways was reported. Gathering all reported results, they discovered a positive-feedback loop embedded in the crosstalk between the Wnt and ERK pathways, and have developed a plausible model that represents the role of this hidden positive-feedback loop in the Wnt/ERK pathway crosstalk based on the integration of experimental reports and employing established basic mathematical models of each pathway. The positive-feedback loop can generate bistability in both the Wnt and the ERK signalling pathways, and this prediction was further validated by experiments.

Regulatory Single-Nucleotide Polymorphisms

The REGULATORY-GENOMICS (2007) project observes that much remains to be learned about the molecular mechanisms that control expression of human genes, and about the variations in gene expression that underlie many pathological states, including cancer. This is caused in part by lack of information about the "second genetic code" – binding specificities of transcription factors (TFs). Deciphering this regulatory code is critical for cancer research, as little is known about the mechanisms by which the known genetic defects induce the transcriptional programmes that control cell proliferation, survival and angiogenesis. In addition, changes in binding of TFs caused by single-nucleotide polymorphisms are likely to be a major factor in many quantitative trait conditions, including familial predisposition to cancer. The project aims to develop novel genomics tools and methods for determination of TF binding specificity. These tools will be used for identification of regulatory single-nucleotide polymorphisms that predispose to colorectal cancer, and for characterisation of downstream target genes that are common to multiple oncogenic TFs. A CIS-modules (2007) database is available containing genome-wide enhancer predictions.

Cell Cycle

Cell Cycle Functional Modules

Deregulation of the cell cycle is primarily responsible for cancer. The two main characteristics of all neoplastic cells are abnormal proliferation and aneuploidy, two direct consequences of cell cycle deregulation. DIAMONDS (2007) has substantially developed the systems biology analysis of the cell cycle. This project has created a firm basis for a high-throughput functional analysis of findings and hypotheses. It analyses

- Disturbances of the cell cycle regulatory network which lie at the basis of many cancer types
- A comparative approach to illuminate the variation in the intrinsic stability of cell cycle controls in plants and animals, leading to new insight into how to combat proliferative disorders
- The mode of action of cell cycle regulators and provides a basis for identification of potential therapeutic targets

Some of the major results are summarised in DIAMONDS-D3.5 (2007). The dynamic activation of the functional modules is key in securing a fully functional daughter cell. To investigate the activation of the functional modules as well as getting a starting point for finding important cell cycle regulatory motifs, microarray expression data from synchronously growing cell cultures of *Saccharomyces cerevisiae*, *Schizosaccharomyces pombe*, *Arabidopsis thaliana* and *Homo sapiens* were analysed with the state-of-the-art analysis method (described with detailed references in deliverable D3.4).

A set of functional modules has been identified and their regulation during the cell cycle has been examined and compared across *H. sapiens*, *S. cerevisiae* and *S. pombe*. This comparison revealed that transcriptional regulation is not conserved on a single gene level, but instead appears to be conserved at the functional modules level, which is a major result for the DIAMONDS project. Furthermore, a set of *cis*-regulatory motifs has been identified in *S. cerevisiae* and a genome-wide search of the presence of these motifs in *S. pombe* and *S. cerevisiae* has been conducted.

Tumour Suppressor Genes

pRb Tumour Suppressor

The EUROHEAR (2007) project's objective is to understand the hearing mechanisms involved in the inner ear and the genetic and molecular mechanisms underlying hearing impairment. Mantela et al. (2005) showed that the tumour suppressor gene pRb and the encoded protein pRb are expressed in differentiating and mature hair cells. In addition to pRb, the cyclin-dependent kinase inhibitor (CKI) p21 is expressed in developing hair cells, suggesting that p21 is an upstream effector of pRb activity. p21, which also participates in the cell cycle, apparently cooperates with other CKIs. EUROHEAR-newsletter (2007) discusses bioinformatics approaches to genomic signal processing. In the same way that electrical engineers developed signal processing techniques to extract and refine signals often buried in confounding noise, so genomic signal processing practitioners have developed techniques to extract information from gene expression experiments.

p53 and Apoptosis

Role of p53

A recent book about p53 (Hainaut and Wiman 2007) describes the central role of p53 in cancer. p53 has emerged as a key guardian that triggers apoptosis, cell death, in the case of cell abnormalities, and is therefore an important target for novel cancer therapy, especially since the p53 gene is mutated in a large fraction of human tumours.

p53 and Mdm2 Feedback Loops

COMBIO (2007) uses systems biology methods to study the disruption of the negative-feedback loop between p53 and Mdm2, which is sufficient to generate stable and active p53, thus targeting tumour cells to cell cycle arrest, senescence or apoptosis. Such an approach provides vital complementary information to more clinically based investigations. An example of the outcome of this research has been discussed (Krull et al. 2006) using TRANSPATH (2007), which is an information resource for storing and visualising signalling pathways and their pathological aberrations, and a database about signal transduction events. It provides information about signalling molecules, their reactions and the pathways these reactions constitute. The representation of signalling molecules is organised in a number of orthogonal hierarchies reflecting the classification of the molecules, their species-specific or generic features and their post-translational modifications. Reactions are similarly hierarchically organised in a three-layer architecture, differentiating between reactions that are evidenced by individual publications, generalisations of these reactions to construct species-independent "reference pathways" and the "semantic projections" of these pathways. A number of search and browse options allow easy access to the database contents, which can be visualised with the tool PathwayBuilder. The module PathoSign adds data about pathologically relevant mutations in signalling components, including their genotypes and phenotypes. TRANSPATH and PathoSign can be used for visualisation and modelling of signal transduction networks and for the analysis of gene expression data.

p53 Mutations

An ambitious project called Mutp53 (2007) aims to develop therapies against mutant p53. The p53 tumour suppressor gene is mutated in almost 50% of all human tumours, including most tumour types. A majority of these mutations are point mutations that give rise to single amino acid substitutions in the so-called core domain, i.e. the central domain of p53 that binds to DNA in a sequence-specific manner. Mutant-p53-carrying tumours often show poor response to conventional anticancer therapy such as radiotherapy and chemotherapy; therefore, novel therapeutic strategies that target mutant-p53-carrying tumours could significantly improve clinical outcome in cancer patients. p53 mutations not only serve to inactivate normal (wild-type) p53, but may also endow the mutant protein with novel properties, so-called gain-of-function activities, that could contribute to tumour development. This project is focused on exploring mutant p53 as a target for novel anticancer therapies. Such therapies should aim to either abrogate the gain-of-function effects of mutant p53, or restore wild-type-like properties to mutant p53, so that it can regain its tumour-suppression capabilities. A multidisciplinary approach will be undertaken to explore and exploit the contribution of mutant p53 to cancer. One component of this project will focus on the molecular properties of mutant p53: structural studies will pinpoint the changes that particular mutations inflict on the structure of p53, and allow the classification of mutant p53 into distinct subclasses. In parallel, biochemical studies will explore the mode of action of mutant p53 within cells, including its impact on patterns of gene expression, identification of specific DNA sequences targeted by mutant p53 and discovery of mutant-p53-interacting cellular proteins. Preclinical models for mutant-p53-driven cancer will also be developed, as a critical instrument for preclinical studies. The other component will aim at translating this wealth of information into better cancer therapy.

p53 Database

A key tool based at one of the partners of Mutp53 (2007) is the IARC TP53 Mutation Database (IARC-tp53 2007; Petitjean et al. 2007), which compiles all TP53 gene mutations identified in human cancers and cell lines that have been reported in the peer-reviewed literature since 1989. The database includes various annotations on the predicted or experimentally assessed functional impact of mutations, clinicopathologic characteristics of tumours and demographics of patients. The following datasets are available:

- TP53 somatic mutations in sporadic cancers
- TP53 germline mutation in familial cancers
- Common TP53 polymorphisms identified in human populations
- Functional properties of P53 mutant proteins
- TP53 gene status in human cell-lines

Apoptosis Modelling

The modelling libraries of ESBIC-D (2007) contain modules relevant for combating diseases, including cell cycle models for the understanding of origin of cancer. Several models for inhibition of apoptosis through different signalling cascades (EMI-CD-APOPTOSIS 2007) have been developed. Apoptosis is a distinct form of cell death that is functionally and morphologically different from necrosis. Nuclear chromatin condensation, cytoplasmic shrinking, dilated endoplasmic reticulum and membrane blebbing characterise apoptosis in general. The two principal pathways of apoptosis are (1) the Bcl-2 inhibitable (mitochondria-mediated or intrinsic) pathway induced by various forms of stress like intracellular damage, developmental cues and external stimuli and (2) the caspase-8/10 dependent (extrinsic) pathway initiated by the engagement of death receptors. The caspase-8/10 dependent or extrinsic pathway is a death receptor mediated mechanism that results in the activation of caspase-8 and/or caspase-10. Activation of death receptors like Fas/CD95, TNFR1 and the TRAIL receptor is promoted by the TNF family of ligands including FASL (APO1L or CD95L), TNF, LT-α , LT-β , CD40L, LIGHT, RANKL, BLYS/BAFF, and APO2L/TRAIL. These ligands are released in response to microbial infection, or as part of the cellular, humoral immunity responses during the formation of lymphoid organs, activation of dendritic cells, stimulation or survival of T, B and natural killer cells, cytotoxic response to viral infection or oncogenic transformation. The Bcl-2 inhibitable (intrinsic) pathway of apoptosis is a stress-inducible process, and acts through the activation of caspase-9 via Apaf-1 and cytochrome *c*. The rupture or permeability of the mitochondrial membrane, a rapid process involving some of the Bcl-2 family proteins, leads to the release into the cytosol of proapoptotic proteins, previously located in the intermembrane space of the mitochondria. Examples of cellular processes that may induce the intrinsic pathway in response to various damage signals include autoreactivity in lymphocytes, cytokine deprivation, calcium flux or cellular damage by cytotoxic drugs like taxol, deprivation of nutrients like glucose and growth factors like epidermal growth factor, with anoikis and transactivation of target genes by tumour suppressors including p53. In many non-immune cells, death signals initiated by the extrinsic pathway are amplified by connections to the intrinsic pathway. The connecting link appears to be the truncated BID protein a proteolytic cleavage product mediated by caspase-8 or other enzymes.

The new FP7 (2007) project APO-SYS will greatly extend apoptosis modelling work.

p53, p63 and p73 Comparisons

EPISTEM (2007) is investigating the role of p63 and related pathways in epithelial stem cell proliferation and differentiation and in rare ectrodactyly ectodermal dysplasia. As part of this activity, it is investigating the regulation and involvement of p63 and related pathways in skin differentiation, the maintenance of the proliferative capacity of epithelial stem cells and the transition of ectodermal cells to epidermal stem cells. It will create a bioinformatics platform with a graphical interface that integrates the separate datasets obtained throughout the project (vertical comparative genomics) with public datasets from other species and the phylogenetic analysis of the p63, p53 and p73 protein family (horizontal comparative genomics) to predict p63, p53 and p73 functional interaction partners.

Cell Immortalisation, Tumourigenesis and Senescence

Irreversible Growth, Apoptosis and Premature Senescence

The ENFIN-wp5.2 (2007) project is investigating the mechanisms of cellular senescence. In normal cells, growth factor signals (like those from TGF- β1) are interpreted by tightly regulated networks of signal-transduction proteins that regulate the appropriate cellular response. Cancer cells are often unresponsive to normal signalling cascades, and thus their responses are very unpredictable. However, there is emerging evidence that acute activation of a single mitogenic oncogene in mammalian cells not only promotes cell proliferation but surprisingly also simultaneously switches on growth-opposing cellular programmes. These include apoptosis and an irreversible growth arrest termed "premature senescence" (Shay and Roninson 2004). Analogous to TGF-β signalling in normal cells, the specific genetic constituency of a cancer cell, which qualitatively and quantitatively determines which programmes dominate, determines whether the given cell divides, stops or dies in response to the oncogenic stimulus. Finding either genetic or pharmacological means to tilt this balance towards cellular senescence or apoptosis provides an intriguing opportunity to selectively target oncogene-expressing cancer cells to destruction. The partial network reconstruction will be used to rank candidates in these pathways for further investigation by experimental means. These candidates will be tested by functional knockdown analysis, and iterative bioinformatics analysis will refine the procedures and statistics for effective partial network reconstruction. Comparative modelling of pathways is used to project the rich knowledge of the yeast mitotic pathway into the mammalian context. Comparative mammalian mitosis predictions will be tested using RNA interference assays developed for this biological area. Genome-wide RNA intereference disruption experiments are planned in the MitoCheck (2007) project. After gene knockdown, mitotic phenotypes are recorded by digital fluorescence video microscopy. Phenotypes are quantitatively scored by automated image processing routines, generating a multiparametric mitotic "phenotypic fingerprint" for each gene. The MitoCheck (2007) project will generate a considerable amount data but there is limited scope for data mining the results, in particular in the context of comparative information. Using the ENFIN (2007) core and analysis layer, it will correlate the phenotypic information from the genome-wide screen with the comparative pathway reconstruction.

Predictive Dynamic Model

The VALAPODYN (2007) project is developing a validated predictive dynamic model of complex intracellular pathways related to cell death and survival. It seeks to further the development of multidisciplinary functional genomics relating to complex biological processes and cellular networks. The project is concerned with both DNA and protein applications, to be followed by innovative dynamic modelling of pathological disease states such as epilepsy and cancer, in order to validate the model. The overall aim is to develop an innovative systems biology approach, in order to model the dynamics of molecular interaction networks related to cell death and survival in the organism.

Multistep Tumourigenesis

Virtual Tumour Progression Features

CVIT (2007) brings together an international group of investigators with interest in the biomedical, the computational and the mathematical aspects of cancer research, and interacts with the Advancing Clinicogenomic Trials on Cancer (ACGT 2007) action called "Technologies and tools for in silico oncology". CVIT's long-term goal is to develop a generic, module-based toolkit for modelling and simulating selected cancer types of interest, such as breast cancer, brain cancer and melanoma, following a complex systems approach. Combined with biomedical data, this modelling toolkit will have significant value for experimental cancer research as it allows researchers to properly study cancer initiation and such critically linked progression features as invasion, angiogenesis and metastasis in the context of an emergent system. Ultimately, this toolkit will also have important clinical applications, including trial design and management as well as patient outcome predictions.

Succession of Genetic Mutations in Colon Cancer

Not all modelling depends on computers. Johnston et al. (2007) have shown the power of mathematical modelling of cell population dynamics in the colonic crypt and in colorectal cancer. Colorectal cancer is initiated in colonic crypts. A succession of genetic mutations or epigenetic changes can lead to homeostasis in the crypt being overcome, and subsequent unbounded growth. They considered the dynamics of a single colorectal crypt by using a compartmental approach which accounts for populations of stem cells, differentiated cells and transit cells. The results showed that an increase in cell renewal, which is equivalent to a failure of programmed cell death or of differentiation, can lead to the growth of cancers.

Genomic Integrity and the Development of Cancer

DNA Repair

DNA Repair (2007) focuses on unravelling mechanisms of DNA damage response and repair, an area relevant to cancer, immunodeficiency, other ageing-related diseases and inborn disorders. The project brings together leading groups with multidisciplinary and complementary expertise to cover all pathways impinging upon genome stability, ranging from molecules to mouse models and human disease. The main objective is to obtain an integrated perception of the individual mechanisms, their complex interplay and biological impact, using approaches ranging from structural biology to systems biology. The pleiotropic effects inherent to the time-dependent erosion of the genome and the complexity of the cellular responses to DNA damage necessitate a comprehensive, multidisciplinary approach, which ranges from molecule to patient. At the level of structural biology and biochemistry, individual components and pathways will be analysed to identify new components and clarify reaction mechanisms. The interplay between pathways and crosstalk with other cellular processes will be explored using both biochemical and cellular assays. To better understand the function and impact of DNA damage response and repair systems in living organisms, the existing unique and extensive collection of models (mutant yeast cells and mice) is used to engineer and analyse new mutants impaired in genome stability. The rapid growth in genomic and proteomic technologies will be exploited to identify novel genes involved in genome surveillance. Bioinformatics and high-throughput systems will be used for analysis of gene expression, proteomics will be used for identifying putative functions of such genes and their proteins, and similar global genome analytical tools will be used to identify interactions with and effects on other cellular processes.

Angiogenesis and Lymphangiogenesis

Angiogenesis

ANGIOTARGETING (2007), a project on tumour angiogenesis research, focuses on the identification of novel genes and gene products that regulate tumour angiogenesis and on validating such products as therapeutic targets. Solid tumour growth depends on a continuous supply of nutrients by new vessels that grow into the tumour. This process, termed “tumour angiogenesis”, is regulated by a number of complex factors involving both tumour and host cells. How the tumours communicate with the normal cells to produce blood vessels has gained increasing attention over the years and it has recently been shown that targeting the host vasculature has a therapeutic potential for certain tumours. Libraries of complementary DNA will be developed from the tumour and normal vascular transcriptome. The data will be processed by the bioinformatics group, where the objective is to provide bioinformatics support and integration of data for the whole project.

Lymphangiogenisis

The lymphatic vasculature is essential for the maintenance of fluid balance in the body, for immune defence and for the uptake of dietary fat. Lymphatic vessels promote metastatic spread of cancer cells to distant organs – a leading cause of death in patients with cancer, and a major obstacle in the design of effective therapies. Lymphangiogenomics (2007) aims to discover novel genes important for lymphatic vascular as opposed to blood vascular development and function and to study the functional role and therapeutic potential of their gene products in lymphangiogenesis. The methods used include large-scale knockout and knockdown of the mouse genome, the embryonic stem cell technology, knockdown of zebrafish (*Danio rerio*) genes by morpholino-antisense and positional cloning of disease-susceptibility genes involved in lymphangiogenesis. These studies will provide fundamental new understanding of the molecular and cellular basis of lymphangiogenesis and therefore enable scientists to develop therapies to suppress the growth of lymphatic vessels (e.g. for cancer, inflammatory diseases) or to stimulate their growth (e.g. for tissue ischemia, lymphedema). Extensive bioinformatics support is being provided to facilitate management and interpretation of data from gene and protein expression profiling and positional cloning. A bioinformatics platform is being developed to integrate the expression data generated within the consortium with public-domain data. Cell type specific genes are identified in a classification procedure based on profile similarity with user-provided reference genes. Major challenges include annotation of gene identities and weighting of different data types. Classification methods span from statistical methods such as Pearson correlation and logistic regression to more complex computational methods such as decision trees and support vector machines. The platform supports identification of cell type specific genes for functional evaluation, and modules of co-expressed genes that provide entry points for gene regulation studies.

Tumour Microenvironment Interactions

In addition to oncogenic mutations that act cell-autonomously, tumour cell growth depends on interactions with the microenvironment. The tumour microenvironment consists of cells of haematopoietic and mesenchymal origin, including inflammatory cells, stem and progenitor cells, fibroblasts, endothelial cells and vascular mural cells. Tumour cell growth is known to depend on the interaction of tumour cells with such stromal cells. For example, a growing tumour needs to recruit normal endothelial and vascular mural cells to form its blood vessels. In addition, tumour cells induce stromal cells to secrete factors that contribute to tumour cell growth and invasion. Stromal-cell-dependent interactions represent an attractive target for cancer therapy, because normal cells are genetically stable, and would not be expected to develop resistance to therapeutic agents. The Tumour-Host Genomics (2007) project aims at studying major signalling pathways in mesenchymal and haematopoietic cells, forming a concerted effort to understand tumour–host interactions, and to identify novel therapeutic targets, entailing development of novel advanced functional genomics instruments, technologies and methods to study tumour–host interactions in cancer, and to apply these techniques to the identification of molecules and processes in normal cells which could be targeted by novel anticancer therapeutic agents. Hallikas et al. (2006) noted that understanding the regulation of human gene expression requires knowledge of the "second genetic code", which consists of the binding specificities of TFs and the combinatorial code by which TF binding sites are assembled to form tissue-specific enhancer elements. Using a novel high-throughput method, they determined the DNA binding specificities of GLIs 1–3, Tcf4 and c-Ets1, which mediate transcriptional responses to the Hedgehog (Hh), Wnt and Ras/MAPK signalling pathways. To identify mammalian enhancer elements regulated by these pathways on a genomic scale, they developed a computational tool, EEL (2007), the enhancer element locator. EEL can be used to identify Hh and Wnt target genes and to predict activated TFs on the basis of changes in gene expression. Predictions validated in transgenic mouse embryos revealed the presence of multiple tissue-specific enhancers in mouse c-Myc and N-Myc genes, which has implications for organ-specific growth control and tumour-type specificity of oncogenes.

Metastasis

Metastasis of Breast Cancer

The BRECOSM (2007) project's objectives are to identify genes, proteins and molecular pathways involved in regulating the metastasis of breast cancer to specific organs. It is using a combination of gene expression profiling, bioinformatic analysis, histology of human breast cancer samples, genetic manipulation of transplantable tumour cells and transgenic mouse technology. In addition to finding new genes, it is analysing to what extent genes already known to play a role in breast cancer metastasis specify to which organs breast tumours metastasise. It will establish how the currently known genes that are associated with breast cancer dissemination and new ones fit together into pathways that regulate organ-specific metastasis. Thiery and Sleeman (2006) discussed how complex networks orchestrate epithelial–mesenchymal transitions and how they may be analysed.

Tumour Immunology and Immunotherapy

Cancer Immunotherapy

ATTACK (2007) focuses upon the development of immune-cell therapies to target cancer, including the use of gene therapy approaches to modify T cells. T cells are part of the immune defence machinery which naturally protects against infections and some cancers. T cells can be used to treat some malignant diseases, but many cancers avoid destruction by the immune system. With use of state-of-the-art technologies to target the T cells by introducing artificial receptors, it is hoped to provide these cells with tumour specificity. See also the end of Chap. 3 for a description of the new FP7 (2007) project SYBILLA: Systems Biology of T-Cell Activation in Health and Disease.

Implementation in the Seventh Framework Programme

Seventh Framework Programme Research Project in the Systems Biology of Cancer

A topic in the first call for proposals FP7-CALL-HEALTH-2007-A (2007) is leading to the following project.

Apoptosis

Developing an integrated in vitro, in vivo and systems biology modelling approach to understanding apoptosis in the context of health and disease. APO-SYS: Apoptosis systems biology applied to cancer and AIDS – An integrated approach of experimental biology, data mining, mathematical modelling, biostatistics, systems engineering and molecular medicine. A Europe-wide consortium of experimental biologists, biomathematicians, biostatisticians, computer scientists and clinical scientists will team up to approach cell death pathways in health and disease, placing particular emphasis on cancer and AIDS. The consortium will create a unique database integrating existing and accumulating knowledge on lethal signal transduction pathways leading to apoptosis or non-apoptotic (necrotic, autophagic, mitotic) cell death, perform data mining to integrate system-wide analyses on cell death (genome, epigenome, transcriptome, proteome, lipidome data), and use high-throughput methods ("omics", ChIP-chip and genome-wide siRNA screens) for the experimental exploration of death pathways in human cell lines in vitro and in relevant disease models (*in vitro* in human cells and in vivo in mice and *Drosophila*). In addition, the consortium will establish mathematical models of lethal pathways to devise algorithms that predict apoptosis susceptibility and resistance, obtain data (genome, transcriptome, proteome, lipidome) on clinical samples (cancer cell lines, cancer tissues, serum, and blood samples) and perform biostatistical analyses on them in order to demonstrate the contribution of apoptotic process in human cancers and AIDS. The consortium will integrate the knowledge into mathematical models for the optimal interpretation of clinical data, aiming at optimal diagnostic and prognostic performance as well as at the identification of possible therapeutic targets for the treatment of cancer and AIDS.

Implications of the New Project

This project builds on a wide range of capacities developed in FP6 (2007), such as the bioinformatics and systems biology tools from BioSapiens (2007), EMBRACE (2007), ENFIN (2007) and EMI-CD (2007). It also builds on the systems biology approach to cancer research developed in the ESBIC-D (2007) pilot project and on the wide range of laboratory tools developed, for example in the proteomics area and many other areas. It demonstrates how the full power of the tools developed in FP6 (2007) in bioinformatics, systems biology and supporting high-throughput tools and infrastructures can be unified to work on the most complex and most relevant health problems. It also spans the full range from basic investigation to applied clinical research and drug development, and demonstrates how all research levels may be combined into a highly effective approach. It may be that this project will point the way to the new paradigm for complex disease research and applications.

Chapter 9
Genetic Variation and Diseases

Abstract This chapter focuses on bioinformatics aspects of genetic variation research. Because of the diversity of this field, an extensive discussion is provided of the issues involved. The status of genetic variation research is discussed, on the basis of the genotype-to-phenotype relationship. The data and analysis challenges are analysed, and currently available database information is summarised. The key problems of data generation, capture and analysis are considered. By an exploration of the possible integration of various initiatives, the way forward in this field is analysed. Current research in the Sixth Framework Programme is outlined, and a major initiative in the Seventh Framework Programme to tackle many aspects of these problems is portrayed.

Introduction

Background

This chapter explores one of the great challenges of modern biology, how to integrate all the resources on germ-line and somatic genetic variation into disease research, by focusing on bioinformatics aspects. Extensive but fragmented work is going on worldwide in this field, involving hundreds of separate databases. Studies of genetic variation in humans and model organisms provide major insights into molecular biology processes, evolution, health and disease patterns (Strachan and Read 2004; Griffiths et al. 2000; Jobling et al. 2004; Epstein 2003). Because of the diversity of this field, this chapter opens with an extensive discussion of the issues involved, and then goes on to describe work already done and a major European Commission initiative just starting, directed at addressing these problems, involving several large collaborative projects.

F.B. Marcus, *Bioinformatics and Systems Biology*,
DOI 10.1007/978-3-540-78353-4,

Call to Action

An editorial (Editorial Nature Genetics 2005) and a paper (Patrinos and Brookes 2005) highlighted the need for coordination and interconnection of databases focusing on human genetic variation and associated phenotype relationships. They noted that the need for coordination in this area has long been recognised, but that so far no effective solutions had been found. The current database, data capture and analysis structures are highly fragmented, and a wide range of analyses are very difficult or suboptimal. There are currently a number of worldwide efforts and discussions on moving towards a unified database, e.g. Variome (2007), but they have all come up against the problem of combining very different data types and research fields into a single database. The problem is sometimes compounded by difficulties with data accessibility, especially where patient confidentiality is involved. Improved access, data validation, curation and analysis would be immensely valuable and allow better utilisation of the billions of euros of expert work in human genetics research involving population and comparative genetics, biobanks, clinical trials and pharmacogenetics.

Genetic Variation Workshop

Genetic Variation Workshop Organisation and Goals

As a result of this situation of fragmented data access, a European Commission workshop was organised, leading to a report by Marcus and Mulligan (2006), which provided several insights into the state of genetic variation research. The goal of the workshop was to describe how to provide a database and analysis structure for much of human and model organism genetics. A means to achieve it in the near future was outlined, by a hierarchy of grid-linked databases and tools.

Editorial Encouragement

Further insights are provided in the editorial comment on this workshop report in Editorial Nature Genetics (2006). Entitled “Jousting for HUGOBase”, the editorial refers to a quote “Using data is popular; contributing data is unpopular” as follows:

This is a typically fresh quote from the official report of the far-sighted Workshop on European Database and Analysis Resources for Research in Human Genetic Variation (Marcus and Mulligan 2006). Held on March 2–3, 2006 in Brussels, this effective workshop brought bioinformaticians together with medical, clinical and biological experts to examine ways to extend existing European projects into an integrated human genome variation database along the lines discussed in our

August 2005 Editorial Nature Genetics (2005). The workshop concluded that Europe has already started most of the projects necessary to the success of such an integrated database (which at the global scale we call HUGOBase). Participants also emphasized that it will be necessary to hold a peer-reviewed competition to identify those coalitions that have the capacity to integrate the results of other data producers. A single database for human variation data is unfeasible because of the diversity of producers, disciplines, funding mechanisms and user needs. Fortunately, Europe is home to a number of bioinformatics grid technologies for linking databases (EMBRACE, 2007), and it is hoped that these will provide the interface through which specialised users of the underlying data will apply their own tools.

Status of Genetic Variation Research

Value of Cross-Linking Data

Until now, no effective strategy for achieving database linking or unification has been formulated. Existing collections of genetic relationships, predominantly from Mendelian single gene variation traits, when supplemented by information from model organisms, have provided many fundamental insights into human biology, at both the body and the cellular levels. Differences between healthy people, and also causes of diseases, have some genetic component. If all aspects of the genetic contribution could be identified, they would lead to advances in biomedical research, as well as furthering the cataloguing of human genotype–phenotype relationships. However, the lack of data integration inhibits many research breakthroughs. An integrated genetic variation catalogue would be an immense boon to bioresearch, in areas such as general understanding of human physiological processes in both health and disease, the ability to analyse populations according to different classifications and the diagnosis and treatment of disease.

Linked Databases

The conclusion of the workshop (Marcus and Mulligan 2006) was that an integrated database and analysis structure for much of human and model organism variation genetics should be achieved by database-linking at a European level, and in the near future, by means of a pragmatic and step-by-step approach. The organising principle of the database network would be the genotype-to-phenotype relationship. This combination spans the whole descriptive range of genetic variation, from single DNA base changes to highly complicated biological and clinical phenotypes and diseases. The workshop participants thought it infeasible to attempt to create a wholly new central database, with associated ontologies and standards. In fact, a wide range of databases already exist within Europe, with preliminary links to important databases elsewhere, e.g. OMIM (2007) and dbSNP (2007) in the USA. These databases and ontologies are sufficient and suitable for forming hubs for interlinking in rather straightforward ways. Database linkage could be accomplished using technologies implemented in existing European Union bioinformatics grid projects and data exchange formats. This linkage should be based on a hierarchical system, with one or two major genetic sequence based genomic databases like Ensembl (2007) and its genome browser software packages acting as a hub, with links to broadly based genome variation databases. There would be further links to the many specialised databases of four main types: locus-specific, disease-specific, population and biobank. These interlinked data should be accessed by a variety of tailored user-friendly interfaces.

Genetic Variation Data Sources

Data in the public domain are required for successful and efficient access. Semi-commercial, commercial (non-public) and medical databases could also be connected with the integrated database system in ways that fully respect either commercial or ethical confidentiality. Many of these databases rely on data in the public domain, and already have arrangements for making some data available to academic researchers. Key genetics research programmes include the Wellcome Trust Case Control Consortium (WTCCC 2007). Other studies of disease-focused association and genetic diversity, conducted at the multipopulation level, would provide the type of data needed to underpin full and correct analysis of many other datasets. Unified and more complete datasets would provide improved opportunities for researchers to study association on a genome-wide level.

New Elements Facilitating Data Integration

To facilitate data linkage, a wide range of new tools have become available, including:

- Integrating data-grid protocols and technologies
- The EMBRACE (2007) bioinformatics grid capabilities, which could be implemented for genetic variation
- Recent experience in integrating databases, e.g. Integr8 (2007)
- Integrated analysis pipelines (BioSapiens 2007; ENFIN 2007)
- Genome browsers (Ensembl 2007)

New high-throughput technologies are becoming available, greatly lowering cost and allowing new data to be more complete. Major scientific support data is being provided from large-scale comparative genomics projects including non-human model organisms, e.g. genotype–phenotype data from European Commission funded mouse projects. Relevant clinical data are increasingly computerised and publicly available. There is a new trend, started in the UK and now also implemented by the NIH (2007) in the USA, towards public release of all data generated in large-scale genetics research projects. In some cases, there is complete public access to control genotype data, and bona fide researcher access to additional data. In some projects, all raw genotype data are released, although there will be some restrictions on initial use. "Big Pharma" is also interested in contributing to these open datasets. In the past, genetic studies of complex diseases have not met with the anticipated success, for example in statistics from human association studies. Most researchers recognise a considerable lack of statistical analysis power and lack of genome coverage for many previous association studies. However, the current generation of association studies has reasonable power and allows genome-wide testing. Very large scale association projects are in progress in the UK (WTCCC 2007) and in the USA with the planned GAIN (2007) project.

Genotype to Phenotype

Biological and Medical Research: Genotype to Phenotype

The guiding organising and scientific principle for linking databases is often the genotype–phenotype relationship, which can involve an extremely detailed and multilevel classification. Simple genotypes (one mutation) and phenotypes (one disease) were the key principles that Victor McKusick used in the USA to found the modern approach to genetics databases 40 years ago, with OMIM (2007).

Genotype

A genotype can be much more complicated than a single single-nucleotide polymorphism (SNP) in a protein coding gene listed in dbSNP (2007), which is integrated with other Entrez-models (2007). The genotype can include SNPs, haplotypes, locus-specific databases (LSDBs), multiple copies, non-coding DNAs, quantitative trait loci (QTL), epigenetic (histones and methylation) and environmental effects, full sample classification and characterisation. SNPs produce a huge number of types and "consequences", with ten million human common variants, and additional minority variants, including non-synonymous, synonymous, untranslated regions, regulatory, GT/AG splice changes, stop gains and frame shifts. Genotype classification needs to look at the context of mutations, including the roles of:

- Abnormal copies and phenotypic differences
- Local DNA sequence context
- Mutation frequency (by type)

- Genomic loci (comparative analysis)
- Mutational spectra (design strategies and comparisons)
- Environmental and population context (providing differential effects of genetics for mutations)

Phenotype

Phenotype (traits or characteristics) information can be used as input to data mining, genetic association, systems biology, physiology and epidemiology analyses. Classifications of phenotype can include normal to altered gene expression, protein–protein interaction, pathway, and the cellular, tissue and organism response. Humans have by far the most complex phenotype classification, developed in medicine in relatively modern terms over the past couple of hundred years. Even medical phenotypes can be strongly subdivided, since clinicians tend to combine and eliminate data, so as to efficiently identify treatment for a global phenotype. To extend and to provide a firmer basis for analysis of data, further studies focusing on gene expression in relation to haplotypes and in duplicated and deleted genomic regions would provide essential data. Such data are not only available from animal models. Patient-derived cell lines provide an enormous resource for such studies (which should be collected and analysed). Other phenotypic variability studies might include:

- Underlying genetic heterogeneity of inherited disorders in populations – linking mutant genotypes to clinical phenotypes
- Interactions between multiple susceptibility factors and environment
- Differentiated and categorised neutral versus pathogenic variations
- Role of sequence variation and modifiers in monogenetic disease
- DNA variation in complex traits
- Health risk with variations associated with particular diseases
- Role of somatic mutations and variations
- Careful correlations of genotype to phenotype
- Outcomes and extended molecular phenotypes, levels of clinical subphenotypes, endophenotypes, e.g. osteoarthritis

Data and Analysis Challenges

Standards and Ontologies

Protocols for standards, ontologies, submission and data exchange are essential for successful data linking. Fortunately, these already exist in very extensive formats, such as GO (2007) at the genetics level and medical classifications at the physiological and pathophysiological levels. Complex XML-based submission and exchange formats have already been specified in extensive detail. Nevertheless, a significant amount of work still remains to be done in terms of fully agreed standards, especially for complex phenotypes. Detailed standards and ontologies will require the following:

- Standard nomenclature of genetic variants.
- Guidelines for contents, database structure and standards.
- International collaborations.
- Control population data across Europe.
- Genetic epidemiology centres working on joint standard operating procedures in a quality control network throughout Europe. Standards for submission and deposition are also crucial and a huge field.

Relevant projects include MolPage (2007), MolTools (2007), EUMORPHIA (2007), and those discussed in the Biobanks (2005) conference. Interconnection would also be helped by developing relevant standards and tools, such as Ensembl (2007), BioMart (2007) and the Polymorphism Markup Language.

Data Submission

Data may be submitted via journals or directly into databases or both. Incentives to laboratory personnel for database deposition should be provided by funders and by publishers of scientific journals. A full range of consultations should be initiated with publishers of journals to investigate ways of improving direct deposition and facilitating data mining via publications standards. In the future, journals and databases may be replaced by "database journals", wherein results are deposited directly into Internet-accessible structured depositories, which are interconnected into a "bioknowledge-web". The genotype–phenotype challenge could encourage this practice. The Web is the easiest and least expensive place to publish work, a fact which would also encourage the inclusion of negative results.

Display and Analysis Tools

To maximise utility and attractiveness of submission via display and analysis tools, a set of principles should be developed and observed, with user-friendliness at the top of the list. User-type-specific front-end interfaces are essential for display, bioinformatics research, association studies and systems biology analysis and simulators. At the bioinformatics level, tools should be developed for association studies across a wide range of genetic data to answer complex queries. Better laboratory-based capture of genotype–phenotype information is essential. Developments are required in the areas of database construction, maintenance and software packages, and a special phenotype vocabulary for LSDBs, national databases and linked databases.

For research geneticists, interfacing via a genome browser is the most attractive means of working, supplemented by data links. On the other hand, clinicians and medical researchers generally do not like to work in genome coordinates. Using data is popular; contributing data is unpopular. This reluctance highlights the importance of having various interfaces suited to user preferences, independent of the internal links between databases. Clinicians would follow the route LSDB → genome (Ensembl) → many links, e.g. OMIM (2007), HGMD (2007). Their requirements include:

- Reliable LSDB interfaces, with links to other (genome) information
- An up-to-date and 100% complete list of known gene variants (including rare variants)
- A reference sequence showing nucleotide numbering for the gene
- A field with information regarding the reported pathogenicity of that variant
- A reference to the source of the information
- Any other tools/links connected to his/her subject of interest

A wish list for all user communities might include:

- An open genotype–phenotype database
- Initial basic functionality (tracking variants which change phenotype)
- Open access (commercial data must be accessible to be integrated)

Analysis Challenges

Analysis of these data presents major challenges. The highest-priority issues include:

- Standardisation and database integration
- The "phenotype data" representation challenge
- Handling association data (software tools for data generators)
- Publication bias to normal results (bring in all study findings, including abnormal results)
- A standard model for classifying DNA variation
- Copy number variation (major genetic effects, complex informatics)
- A generic phenotype data model
- A prototype genetic association database
- Convenient database applications for genotyping laboratories
- Data submission tools for genotype–phenotype data

Linking to Systems Biology Analysis

Systems biology analysis requires the type of data input used by PyBioS (2007), which simulates a wide range of metabolic and expression pathways for healthy and disease states. The data include substrates and products of a reaction, stoichiometry, catalysing enzyme, kinetics, reactants and enzyme concentrations. The program also links to several databases and tools: KEGG (2007), Reactome (2007), TRANSPATH (Krull et al. 2006), SRS (2007), BioCyc (2007), Kinetikon (2007) and "raw" experimental data (expression data, protein–protein interaction). To analyse the effects of genetic variations, it is necessary to know how they activate or deactivate certain key pathways.

Appropriate links to network and pathway databases and quantitative data on the effects of genetic variations are required. A set of such databases have been developed by BioBase (2007), dealing with different aspects of gene regulation (TRANSFAC, TRANSCompel, TRANSPro, TiProD), protein–protein interactions of whole proteomes (HumanPSD) and signal transduction (TRANSPATH) for intercellular (specifically endocrine) signalling networks (EndoNet). They are complemented by databases on pathologically relevant mutations of genes encoding regulatory proteins (PathoDB, PathoSign) and disease-involvement of human proteins (HumanPSD/Disease Reports). Together they constitute an information infrastructure useful for projects that link genotype data to molecular and clinical phenotype information.

Databases

Databases to Be Linked

A mixture of databases and associated analysis tools is currently available worldwide, with many in Europe. Some have been developed in publicly funded projects or with institutional budgets, are still maintained, and are fully in the public domain. Others have been transferred to commercial exploitation to generate revenue for maintenance and upgrading. Still others have been generated as purely commercial products, nevertheless providing some of their contents and services free of charge for users from non-profit entities. A range of mechanisms to balance the interests of academic researchers with those of commercial vendors have been implemented, and are often operational and sustainable.

In general, there are many aspects to consider in choosing databases to be linked:

- Different genetic sources in humans and model organisms
- Different types of data
- Raw versus derived data
- The fact that all genotype–phenotype data are not equally useful
- Compilations of all associations

Databases can be characterised as follows:

1. *Large public databases*, partly linked by periodic data exchange and hot-links include Ensembl (2007), dbSNP (2007), OMIM (2007) and UniProt (2007).

Ensembl (2007) already handles a wide variety of variation data. Variations can be SNPs, and "reasonable" indels (insertions and deletions). It can handle multiple sources, genotypes in multiple populations, and both heterozygote (human) and strain-based (mouse and rat) scenarios. Ensembl is very flexible, allowing for stable and sensible handling of variation. It has the ability to handle larger genome polymorphisms and resequencing data, and to scale to thousands of people and millions of genotypes. This allows integration with functional and comparative genomics data.

2. *Medium-sized general databases* include public and semipublic (charges to commercial customers) databases.

   - *The Human Gene Mutation Database* (HGMD 2007) represents a comprehensive core collection of data on germ-line mutations in nuclear genes underlying or associated with human inherited disease. Data catalogued include single base-pair substitutions in coding, regulatory and splicing relevant regions, microdeletions and microinsertions, indels, triplet repeat expansions, gross deletions, insertions and duplications, and complex rearrangements. Human gene mutation is an inherently non-random process. The nature, frequency and location of the mutational lesion are all strongly influenced by the local DNA sequence context. HGMD may be exploited to study the role of the local DNA sequence environment (e.g. repetitive sequence elements, sequence homologies and specific motifs) in mediating mutational events and to explore the nature of the underlying mechanisms. HGMD provides the only comprehensive collection of data on human gene mutations causing inherited disease and as such provides a key means of linking mutant genotypes to clinical phenotypes. Since functional SNPs with or without known disease relevance are also included, there is already a natural bridge between the pathological mutations in HGMD and the predominantly neutral SNPs catalogued in other databases.
   - *HGVbase* (2007) focuses on all forms of variation and any association studies that connect such variants to any phenotype. This includes published and (mostly) unpublished data. In practice, it will mainly capture genetic association evidence between DNA variants and complex disease, where the genetic component alters risk but does not cause the disease (neither necessary nor sufficient to account for the observed phenotype). It will also capture environment data – the other major contributor of complex disease causation. The records will be extensive in scope, carrying detailed phenotype, sample, population, assay, genotype, allele, haplotype, marker and sequence data, along with concluded *p* values for disease associations (single point effects and synergistic interactions) plus citations, free text and keyword information. Genome annotation (e.g. exon, coding sequence, splice sites, repeats) will also be available for guiding database searches, along with a range of submitter information.

3. *Predominantly commercial databases* include those available through DeCode (2007) Celera (2007) and BioBase (2007).

- *DeCode* (2007) is an impressive resource, combining genealogy, phenotype and genetics. Many potential discoveries are feasible by mining such data, although many challenges remain.

4. *Specialised LSDBs and disease-specific, population and ethnic genetic databases* include COSMIC (2007) and p53 data. Somatic mutations are well handled by COSMIC. Many other databases exist, such as TRANSFAC (2007) for transcription factors and LOVD (2007). Different population databases to assess heterogeneity based on ethnicity and future databases should be linked. There will be a role for many such databases to capture the full spectrum and scale of association studies being conducted globally. It is critical to capture all or most data to distinguish true from false (chance) positives, since it is unreliable to rely on published findings alone. Hence, reliance is preferable on databases like HGVbase (2007) that gather primary data (some with specific focuses such as cancer, nations and pathways), plus interfaces allowing searches in multiple databases. Interface design and standards are essential. Central browsers such as Ensembl (2007) may only be able to (and should only) include summary-level information from these many "association databases"' (i.e. markers, phenotype name and p values) for presentation graphically, with links back to association databases.
5. *Cancer-related databases* include both somatic variations as present in COSMIC (2007) and TCGA (2007), and germ-line variations such as in IARC-tp53 (2007). Again as in Chap. 8, cancer is a special area owing to the huge amount of data generated in this area. Cancer-related data present special challenges in this respect.
6. *Genotype–phenotype raw data archives* are also required. A strategy is required to harmonise raw data collation and annotation, and for tools to store and disseminate the data. This is non-trivial, since some file sizes from single studies are tens of terabytes. Analysis tools for statistical genetics are essential, since they are not as developed as other bioinformatics areas in standardisation and capability.

Approaches to Linking Databases in the Public Domain

By concentrating on linking data which are already in the public domain, one devolves all issues concerning access, patient privacy and confidentiality to the local level. The issues are resolved at the level of each institution (e.g. hospital) or government (state or country), each with its own ethics committee, legislation, medical procedures and traditions. Links to "restricted" databases can be developed on a case-by-case basis, keeping in mind overall goals of maximising public access for research, where appropriate. Discussions indicate that these database owners feel that there are solutions to making data publicly available, while protecting their value-added commercial viability or medical confidentiality.

Data Access

The question of data access and property rights to biobanks is highly controversial. Biobank initiatives in Iceland, Estonia and the UK propose the policy of no rights of individual donors or patients to control use of their tissue, by implementing a blanket consent. This controversy is an illustration of why it is important to concentrate on data that have already been put into the public domain, and at a local level, based on local ethical and consent policies.

European-Level Support and International Collaboration

Within the European Union and its Research Framework Programme, including associated states, there is a full enough range of databases (including those developed as international collaborations) to form a fully functional and valuable set of software and databases of major value. A European initiative could forge workable links with its US and other counterparts, via extended grid technology. The Europe Union could also lead as a data provider, because of excellent annotation of data and record keeping in the health care systems.

National-Level Programmes

National-level funders, such as the Wellcome Trust (which although a charitable trust operating internationally, it has the breadth of a large national programme), already have wide-ranging activities in data sharing and databases. They support major initiatives to generate large-scale datasets for the research community: e.g. the genome sequencing projects, and the Structural Genomics Consortium. They also support activities specifically related to human genetic variation, e.g. the SNP Consortium, the International HapMap Project, the Case Control Consortium, the Cancer Genome Project and research in genomic structural variation. National bodies have a key role in simultaneously supporting local resources and facilitating external collaborations. Coordinated approaches are essential for funding of databases, curators, sustainability and access, in a variety of academic/commercial environments.

Role of Model Organism Databases

The role and importance of model organisms is vital. All life on earth is linked by evolution. Even the most basic organisms provide relevant genotype–phenotype relationships in pathways and interactions conserved through evolution. Such data are often not available from humans, for a variety of experimental and ethical reasons. For example, lethal germ-line mutations, e.g. inherited homozygous or mutated egg or sperm heterozygous mutations, lead to death soon after fertilisation and yield no population data. Recently, phenotype categories have been vastly expanded and characterised in model organisms such as mouse and zebrafish. The best genotype–phenotype data for certain kinds of human genetic inferences are via mouse, fly and worm (not human!). In these model organisms, studies of large-scale variation aspects are well advanced.

Data Generation

Related Genetics Research and Infrastructures: Biobanks and Testing

A number of related and integrated genetics research programmes are required to provide the type of data needed to underpin full and correct analysis of linked database resources. These programmes have been discussed in conferences and biobanks workshops (Biobanks 2005). It was concluded that activities should include very extensive studies of comparative, developmental and functional genomics in human and model organisms as appropriate, which provide both variation data and the biological knowledge underpinning analysis. All activities should be carried out in close and continuous interactive collaboration with data providers and users, including biologists, geneticists, medical researchers and clinicians.

Biobanks and related genetic testing form a key part of the foundations for future health care, enabling:

- Outcome research for individuals carrying risk genotypes
- Prospective follow-up of entire populations
- Detection of subclinical manifestations
- Genotype-based prevention
- Clinical trials to establish procedures
- New algorithms for genotype-balanced randomisation

To develop these possibilities, there are a number of necessary actions required:

- Establish a network of population-representative biobanks that share elements of standardisation
- Establish a network of genotyping centres that are highly standardised
- Create accessibility to these networks for clinicians and clinical expertise
- Provide background genotype frequencies to clinical projects
- Create a repository for all genotypes generated, with tags back into the originating (DNA) biobanks

Control Populations

A key area is the requirement for a control population genetics study, which could be implemented as a mixture of national and European Union collaborative projects. There are also important advantages in combining data sources for disease studies, involving linkage screening, fine mapping and whole genome association. Combined sources would enable very large scale epidemiological studies (multi-centre studies). It would be useful to have a standard set of population samples from across Europe, allowing the identification of a set of markers which can capture most of the allele frequency variation. A common set of guidelines for data release across Europe would facilitate data combination.

Copy Number Variation

To further investigate genome variation of human populations, a database for all copy number variation data (in patients and healthy people) is required, coupled with studies to catalogue this variability. Currently there is no such database, and clinical diagnostic laboratories struggle with the information coming out of genome-wide copy number variation studies in relation to genetic disease. Since clinicians focus on specific pathogenic conditions, an effort to analyse a large set of controls to catalogue the non-pathogenic variations would be very worthwhile.

Patient as Data Source

Discovery of new scientific knowledge is possible from large databases of measurements, observations and interpretations from population-and biobank-based research, by using the patient as a data source, in the sense of accidental experiments. This has the advantage of avoiding sampling bias, which can occur when specific diseases or ethnic populations are chosen. Information is also obtainable about children, which is never the case in clinical trials.

Data-Taking Procedures

Data-taking procedures require improvement, including samples prepared/stored and ready to be rapidly assessed, prioritised marker selection, data analysis and results, so as to satisfy critical statistical issues in complex disease gene identification.

Genetic Etiology

Understanding the genetic causes will further approaches for dealing with complex diseases. With some exceptions, understanding of genetic causes in isolation from environmental considerations will not lead to new therapies and will have little direct impact on the health of European populations with chronic diseases. An important coming challenge will be maintenance of health rather than cure of disease. Molecular prevention will become a major approach and most likely will involve nutrition-based strategies. These remedies will often be applicable to genetic and non-genetic diseases Understanding will require approaches to biobanking and data from the following sources:

- Biobanking data of the consequences of environmental triggers or markers thereof
- Open networks that allows clinicians to integrate information, and not merely to hand over samples
- Internal biobank governance by project officers
- Decentralised systems with centralised inventories and standards

This level of data and supporting procedures will be essential to analyse complex diseases and conditions, such as diabetes mellitus, metabolic syndrome, arterial hypertension, arteriosclerosis, coronary heart disease; hyperlipidemia, hyperhomocysteinemia, rheumatoid arthritis/osteoarthritis, depression/bipolar disease, schizophrenia, Alzheimer's disease, dementia, multiple sclerosis, bronchial asthma, atopic eczema, sarcoidosis, psoriasis, periodontitis, malignant diseases, Crohn's disease, ulcerative colitis and longevity.

The Way Forward

Obstacles

Perceived obstacles to unification of genetic variation data can be largely overcome, by means of a pragmatic and step-by-step approach. A single large database with a single interface is not feasible, the main reasons being the extreme diversity of data producers, scientific areas, funding mechanisms, requirements of users and many incompatible databases that could not be combined. Many data are not accessible owing to "commercial" aspects, data confidentiality, ethical questions and lack of incentives to submit data. Many people consider that genetic variation is too broad a field to be unified, especially with data of highly variable quality.

Linking as a Solution

Rather than develop a unified database and database protocols from scratch, most of the goals of a single database can be achieved in a flexible and useful way by a hierarchy of linked databases that currently exist. Extensive linking capabilities and grid experience have already been put in place by European Union projects relying on data in the public domain. An important goal would be to achieve effective linkage between datasets, which would enable scientific results to be integrated across the entire corpus. Given the huge scale and diversity of the subject area, there will be a role for many "data warehouses" that summarise information and discoveries, probably each with different domains of interest or focus (such as general genetic variation, cancer, cardiovascular disease, published studies, specific populations and pharmacogenomics). The future for research in these areas will depend upon data interconnection for transparent cross-database searching, aided by relevant tools and standards. The large amount of data in the public domain makes linking possible.

Hub Software

The unifying hub program, based on genotype identification, would probably be an existing genome browser program and database like Ensembl (2007), with links to dbSNP (2007) and UniProt (2007) databases, using the distributed annotation system (DAS 2007) and grid capabilities. Development of the EMBRACE (2007) life sciences grid is well advanced. The linked genetic variation databases would be at the level of HGMD (2007) and HGVbase (2007). The further levels of linking would include disease, LSDBs and population genetics databases, e.g. COSMIC (2007) and DeCode (2007).

Customised Entry Points

Because the databases are linked rather than unified, there should be several entry points, and several user-tailored interfaces. Even though a genome browser might act as a hub for data deposition, exchange and communications, and as an entry point for researchers in fundamental genomics, a researcher in clinical medicine should access the data via an interface specialised for LSDBs or disease-specific databases.

Principal Conclusions on Linking Genetic Variation Databases

The way forward may be summarised as follows:

- An integrated database and analysis structure for much of human and model organism variation genetics could be achieved by database-linking at a European level by means of a pragmatic and step-by-step approach.

- The organising principle of the database network would be the genotype–phenotype relationship. This combination spans the whole descriptive range of genetic variation, from single DNA base changes to highly complicated biological and clinical phenotypes and diseases.
- Database linkage could be accomplished using technologies implemented in existing European Union bioinformatics grid projects and data exchange formats. This linkage could be based on a hierarchical system, with one or two major genetic sequence based databases like Ensembl (2007) and its genome browser software packages acting as a hub, with links to broadly based genome variation databases. There would be further links to the many specialised databases of four main types: locus-specific, disease-specific, population and biobank. These interlinked data could be accessed by a variety of tailored user-friendly interfaces.
- Data in the public domain are required for successful and efficient access. Semicommercial and commercial (non-public) databases could also be connected with the integrated database system with variable degrees of access.
- Links could be encouraged with genetics/genomics and disease-oriented research programmes. Targeted research is also required. The study of disease-focused association and genetic diversity, conducted at the multipopulation level, would provide the type of data needed to underpin full and correct analysis of many other datasets.

Research in the Fifth Framework Programme and the Sixth Framework Programme

Population Genetics

The GenomEUtwin (2007) project was a major Fifth Framework Programme initiative in population genetics. European populations and epidemiological cohorts are of special significance in genomic research aiming to characterise the background of common human diseases. The genome sequence, detailed information of genetic variations between individuals, high-throughput molecular technologies and novel statistical strategies create new possibilities to define genetic and life-style risk factors behind common health problems. Studies of large population cohorts are needed to transform the genetic information to detailed understanding of the predisposing factors in diseases affecting most human populations. European twin cohorts provide a unique competitive advantage for investigations of the role of genetics and environment or life style in the origin of common diseases. This project applied and developed molecular and statistical strategies to analyse unique European twin and other population cohorts to define and characterise the genetic, environmental and life-style components in the background of health problems.

Down's Syndrome and Relevant Genetic Regions

Down's syndrome involves trisomy in chromosome 21 and is the most common chromosomal viable abnormality (one in approximately 700 births), and the most common known cause of mental retardation. The extra chromosome 21 is usually the result of maternal non-disjunction in meiosis. Little is known concerning the origin of the different phenotypes of Down's syndrome. The in-depth exploration of the chromosome 21 genome and function greatly contributes to the understanding of the molecular pathogenesis of these phenotypes that additionally include cognitive impairment, dysmorphic features, developmental abnormalities of heart, predisposition to leukaemia and numerous other phenotypic characteristics. In research on Down's syndrome (BioSapiens-WP16 2007), the objective was to improve understanding of trisomy 21, by assembling information on the genomic content and potential function of the different elements of chromosome 21, obtained using the most up-to date bioinformatics methods. This will help in formulating and testing hypotheses to understand why three copies of normal genes result in abnormal phenotypes, which gene products are dosage-sensitive and what accounts for the variability of the phenotypes among trisomy 21 individuals. The results were summarised in the following:

- De16.1 document describing the needs and available resources, both from the computational and the experimental side, for Down's syndrome based on the presentations and discussions at the symposium "Molecular and computational biology of Down's syndrome"
- De16.2 "Annotation of gene products of chromosome 21 at the appropriate level of detail"
- De16.3 "Annotation of conserved non-genic sequences on chromosome 21"
- De16.4 "Annotation of genetic variation of chromosome 21"

As described in BioSapiens (2007) deliverable De16.2, in order to unravel the HSA21 genome content, advantage is taken of participation with the ENCODE (2007) project, which provides a detailed map of genes, exons, transcripts of unknown functions, promoters, enhancers, repressors/silencers, origins of replication, sites of replication termination, transcription factor binding sites, methylation sites, deoxyribonuclease I (DNase I) hypersensitive sites, chromatin modifications and multispecies conserved sequences of yet unknown function in 30 Mb of the human genome. For HSA21 there are two ENCODE regions:

- **ENm005** (32668236..34364221 bp from the telomere) about 1.7 Mb gene-rich manually selected regions
- **ENr133** (39244466..39744466 bp from the telomere) about 0.5 Mb randomly selected region

The orthologous sequences are investigated in rat, chimp, mouse, chicken and dog of ENCODE ENm005. The results are used as a model annotation for chromosome 21. As with other genomic regions, the annotations are integrated into BioSapiens (2007). It would be useful to reach the same level of annotation for the entire chromosome 21, to enhance understanding of trisomy 21 (Down's syndrome).

Experimental Tools for Studying Genetic Variation

MolPage (2007) aims to tackle diabetes and one of its major complications, vascular disease, through the development and application of a range of genomic, proteomic and metabolomic technology platforms to carry out "molecular phenotyping" on a medium to epidemiological scale. The research programme has become possible because of recent developments in "omic" techniques which have allowed researchers to gain unprecedented detailed information on biomarkers (genes, proteins and other molecules) affecting the progression and treatment of common disease. The concept of the MolPage project is to develop "omic" technology tools to permit high-throughput analyses of significant numbers of samples, to ensure that protocols are established for sample collection and storage, and that systems are in place to capture, warehouse, analyse and integrate the range of biological data that will emerge from these studies. In the first instance, "molecular phenotyping" technologies from the project will be applied to biomarker discovery and typing in metabolic disease. A goal is to identify biomarkers that are able to highlight individuals likely to suffer from diabetes and vascular disease.

MolTools (2007) aims to establish genome analysis technologies to monitor extensive molecular repertoires, and with the capacity to investigate even single molecules. Molecular technologies are in a very rapid state of development, the scope for improvement is extreme, and methods are clearly rate-limiting for the progress of biology and biotechnology generally. To study DNA resequencing in order to detect unknown sequence changes, the project is devising methods for rapid, inexpensive analyses of large parts of, and ultimately total genomes. Given sufficiently efficient methods it should be possible to identify genetic variations in DNA sequence that predispose to disease or to desirable properties in domestic animals or plants. Two principal methods to detect unknown sequence variants are investigated: oligonucleotide fingerprinting and mass-spectrometric analyses of transcribed and fragmented amplification products. DNA arrays for genotyping are used to investigate the several million common DNA genetic sequence variants which are known for the human and other genomes, potentially allowing the inheritance to be traced for factors that predispose to common diseases. Unfortunately, current methods have insufficient throughput and precision to study diseases with complex inheritance. Several key limiting factors for such analyses are investigated, including design and synthesis of reagents, parallel analyses of large sets of markers, and array-based readout. These technologies are expected to enable whole-genome association studies for disease gene mapping. Advanced array-based transcriptome analyses are used to measure expression levels of transcripts. Recent clinical studies of cancer and cardiovascular diseases demonstrate that gene expression profiles of large gene sets can identify molecular profiles correlated to disease states, which can then be developed to diagnostic tools. Such tools are likely to enter routine clinical laboratories in the near future. However, established microarray technologies show inherent limitations and drawbacks in terms of cost and sensitivity, and still capture only a fraction of the information represented by expression patterns in tissue. Allelic differences, splice variants and weakly expressed genes are frequently ignored, and it is currently impossible to combine the precision and dynamic range of real-time PCR methods with the extensive parallelism of hybridisation arrays. The aim is to develop means to measure gene expression with the ability to detect allele-specific and splice-variant-specific expression profiles at a strong cost reduction and increase to very high throughput.

Genetic Variation Mapping Data

The goal of the genetic variation work in BioSapiens-WP4 (2007) is to integrate genetic mapping data for complex disease in humans and in animal models with the Ensembl (2007) human, mouse and rat genome browsers, in order to identify functional candidate genes within regions identified by genetic methods, and hence be able to make complex queries into the other work packages' databases. There are many mapping projects under way to identify genes associated with complex disease (such as asthma, diabetes (Zeggini et al. 2007), heart disease and cancer), but so far few genes have been proven to be unambiguously associated. However, often linkage to genomic regions has been established, and in some cases to segments small enough (of the order of a megabase) that computational analysis of the underlying sequence in the region is helpful. Consequently, there is a need for software tools that can project genetic mapping data, including haplotypes and disease linkage and association scores, onto the annotated genome sequence to identify regulatory elements (such as transcription-factor binding sites and splicing enhancers) and genes, and to mine information about gene function (from several other BioSapiens 2007 work areas, from the literature and by computational prediction based on domain composition and protein features, for instance). Ensembl (2007) provides a natural conduit for this purpose.

QTL Data

A Wellcome Trust (2007) programme, whose efforts have been leveraged by BioSapiens (2007), mapped genes at high resolution in the mouse for approximately 20 complex QTL behavioural traits and medically important phenotypes such as diabetes and asthma. The experiment exploits a heterogeneous stock, an outbred population of mice derived from a multigeneration intercross of eight inbred mouse strains. Complete phenotype data were developed on 2,000 animals and genotype data from a genome scan employing more than 3,000 markers. This experiment produced a rich dataset of associations between quantitative traits and a large number of megabase-sized candidate regions. Other groups have extensive patient collections for complex disease, including diabetes, asthma, heart disease and psychiatric disease, and susceptibility to infection by malaria, TB, etc. These projects may ultimately yield data which can be integrated with the human genome browser.

Software Tools

This dataset was used as a model to develop software tools to:

- Map genetic association data and mouse inbred strain haplotype data onto Ensembl (2007), including visualisation methods
- Perform complex automatic queries in Ensembl (2007) to characterise candidate regions of the mouse genome

These tools were generalised for use in any genetic mapping project, based on human, rat or other genome data. QTL mapping experiments were also carried out in the rat to find genes associated with diabetes, using congenic rats, microarray data and metabolomic analyses. These data were integrated in a similar way on the rat genome browser.

Next Steps

The plan (BioSapiens-WP103/110 2007) is to use phenotype data on 2,000 animals and genotype data from a genome scan employing 13,800 SNPs. The dataset is described and the results visualised at GSCAN (2007) and in Valdar et al. (2006). This experiment produced a rich dataset of associations between quantitative traits allowing the identification of 843 QTL for 100 phenotypes. On average each QTL contains 28 genes. Identifying which of these genes is functional for the phenotype is one of the main aims of BioSapiens-WP103/110 (2007). The initial data to be provided comprise a list of the genes classified by QTL, phenotype and disease type. Later on, gene expression data for brain, liver and lung will be provided. This would provide additional information about tissue expression, gene co-expression and which transcripts have expression QTL. These data will be used as a model of a BioSapiens (2007) collaborative gene annotation project. The combined results of the analysis will be collated and then made available via the GSCANDB (2007) Web browser and via DAS (2007) annotation tracks on the Ensembl (2007) genome browser.

A Major Seventh Framework Programme Initiative in Genetic Variation

Projects in Genetic Variation Research

The workshop report (Marcus and Mulligan 2006) provided important input to the European Commission consultation process, which led to a series of related topics in the first call for proposals (FP7-CALL-HEALTH-2007-A 2007). Several proposals were received and evaluated, and the following were chosen for contract negotiation within the available budget. On the basis of past experience, it is very probable although not certain that these proposals will become operating projects, at which time more details will be available in the FP7 (2007) projects catalogue and from the project websites, accessible by searching the Internet for their acronym. The following is the subject (in bold) and the topic published by the Commission (in italics), followed by The project acronym (in bold) and the project abstracts, which are published on the FP7 (2007) projects website.

Genetic variation linked databases: *Unifying human and model organism genetic variation databases.*

GEN2PHEN – Genotype-To-Phenotype Databases: A Holistic Solution. The GEN2PHEN project aims to unify human and model organism genetic variation databases towards increasingly holistic views into Genotype-To-Phenotype (G2P) data, and to link this system into other biomedical knowledge sources via genome browser functionality. The project will establish the technological building-blocks needed for the evolution of today's diverse G2P databases into a future seamless G2P biomedical knowledge environment. The project will then utilise these elements to construct an operational first-version of that knowledge environment, by the projects end. This will consist of a European-centred but globally-networked hierarchy of bioinformatics GRID-linked databases, tools and standards, all tied into the Ensembl genome browser. The project has the following specific objectives: 1) To analyse the G2P field and thus determine emerging needs and practices; 2) To develop key standards for the G2P database field; 3) To create generic database components, services, and integration infrastructures for the G2P database domain; 4) To create search modalities and data presentation solutions for G2P knowledge; 5) To facilitate the process of populating G2P databases; 6) To build a major G2P internet portal; 7) To deploy GEN2PHEN solutions to the community; 8) To address system durability and long-term financing; 9) To undertake a whole-system utility and validation pilot study.

DNA sequencing: *Groundbreaking techniques for DNA sequencing and genotyping.*

READNA, REvolutionary Approaches and Devices for Nucleic Acid Analysis. The READNA consortium is composed of researchers from 12 academic institutions, 3 SMEs and 4 large companies. The goals of the READNA consortium are to revolutionize nucleic acid analysis methods, by 1) improving elements necessary to use the currently emerging generation of nucleic acid sequencers in a meaningful and accessible way, 2) providing methods that allow in situ nucleic acid analysis and methods capable of selectively characterizing mutant DNA in a high background of wildtype DNA, 3) combining RNA and DNA analysis in a single analytical device, 4) providing technology to efficiently analyze DNA methylation (genome-wide, with high resolution and in its long-range context), 5) implementing novel concepts for high-throughput HLA-screening, 6) developing fully integrated solutions for mutational screening of small target regions (such as for screening newborns for cystic fibrosis mutations), 7) developing a device for screening multiple target regions with high accuracy, and 8) implementing strategies for effective and high-resolution genotyping of copy number variations. An important part of READNA is dedicated to the development of the next generation of nucleic analysis devices on individual DNA molecules by stretching out nucleic acid molecules in nanosystems, using alpha-haemolysing nanopores and carbon nanotubes. These approaches will benefit from improved interrogation and detection strategies which we will develop. Their methods and devices will boost the possibilities of genetic research by closing in on the target of 1000 Euros for the sequence of a complete human genome, while at the same time leading a revolution in cost-effective, non-invasive early screening for diseases such as cancer.

Epidemiology: *Molecular epidemiological studies in existing well characterised European (and/or other) population cohorts.*

ENGAGE, European Network for Genetic and Genomic Epidemiology, has, as its central objective, the translation of the wealth of data emerging from large-scale research efforts in molecular epidemiology into information of direct relevance to future advances in clinical medicine. ENGAGE will do this through the integration of very large-scale genetic and phenotypic data already available from a substantial number of large and well-characterised European (and other) sample sets of various types. The initial focus will be an integrated analysis of >80,000 genome-wide association scans available to the consortium, thereby identifying the large number of novel disease-susceptibility variants undetectable in individual studies. Early studies will concentrate on metabolic and cardiovascular phenotypes, with subsequent expansion to apply the methods developed and lessons learned in other disease areas. The ENGAGE framework has been designed to be adaptable to advances that enable global analyses of other sources of genomic variation (e.g. structural and epigenetic variants), and to broadening of the phenotypic spectrum (to genomic endophenotypes in particular). The clinical and public health relevance of the novel disease- and trait-susceptibility variants we identify will be evaluated using the breadth and diversity of ENGAGE cohorts (DNAs and serum/plasma samples from over 600,000 individuals). The final step will be to effect responsible clinical translation of their major findings. As well as advances in the understanding of disease pathogenesis which may underpin novel therapeutic advances, they expect to provide clear proof-of-principle that genetic and genomic discoveries can be translated into diagnostic indicators for common diseases with the capacity to stratify risk, monitor disease progression and predict and monitor therapeutic response.

Epidemiology: *Molecular epidemiological studies in existing well characterised European (and/or other) population cohorts.*

HYPERGENES, European Network for Genetic-Epidemiological Studies: building a method to dissect complex genetic traits, using essential hypertension as a disease model. The project HYPERGENES is focused on the definition of a comprehensive genetic epidemiological model of complex traits like Essential Hypertension (EH) and intermediate phenotypes of hypertension dependent/associated Target Organ Damages (TOD). To identify the common genetic variants relevant for the pathogenesis of EH and TODs, they will perform a Whole Genome Association (WGA) study of 4000 subjects recruited from historical well-characterizd European cohorts. Genotyping will be done with the Illumina Human 1M BeadChip. Well-established multi-variate techniques and innovative genomic analyses through machine learning techniques will be used for the WGA investigations. Using machine learning approaches they aim at developing a disease model of EH integrating the available information on EH and TOD with relevant validated pathways and genetic/environmental information to mimic the clinician's recognition pattern of EH/TOD and their causes in an individual patient. Their statistical design is with two samples run in parallel, each with 1,000 cases and 1,000 controls, followed by a replication/joint analysis. This design is more powerful than replication alone and allows also a formal testing of the potential heterogeneity of findings compared to a single step (one large sample) design. The results represent the source to build a customized and inexpensive genetic diagnostic chip that can be validated in their existing cohorts (n = 12,000 subjects). HYPERGENES is in the unique position to propose a ground-breaking project, improving the methodology of genetic epidemiology of chronic complex diseases that have a high prevalence among EU populations. Designing a comprehensive genetic epidemiological model of complex traits will also help to translate genetic findings into improved diagnostic accuracy and new strategies for early detection, prevention and eventually personalised treatment of a complex trait. The ultimate goal will be to promote the quality of life of EU populations.

Metagenome: *Characterisation and variability of the microbial communities in the Human Body.*

METAHIT, Metagenomics of the Human Intestinal Tract. A detailed understanding of human biology will require not only knowledge of the human genome but also of the human metagenome, defined here as the ensemble of the genomes of human-associated micro-organisms. The METAHIT proposal focuses on the microorganisms of the gut, which are particularly abundant and complex and have an important role for human health and well-being. They will implement and integrate the following activities: (i) creation of a reference set of genes and genomes of intestinal microbes, using high fidelity metagenomic sequencing and full genome sequencing of selected bacterial species; (ii) creation of the generic tools, based on the high density DNA arrays and novel ultra-high throughput re-sequencing techniques, to study the variation of human gut microbiota; (iii) use of the tools to search for correlations between the genes present in the gut microbiota and disease, focusing on the inflammatory bowel disease and obesity, the two pathologies of increasing social relevance in Europe; (iv) study of the genes correlated with the disease, both in terms of their function in microbes and their effect on the host, with the focus on host-microbe interactions; (v) development of an informatics resource to store and organize the heterogeneous information generated within the project, such as gene and genome sequences, gene frequencies in healthy and sick individuals or gene functions and also enriched by information relevant to human gut microbiota from the outside of the project; (vi) creation of the bioinformatics tools to carry out the meta-analysis of the information; (vii) creation of an interface with the stakeholders, including an international board to promote cooperation and coordination in the human metagenome field, and general public. Their project will place Europe in a leading position in this field and open avenues to modulate human gut microbiota in a reasoned way, enabling to optimize the health and wellbeing of any individual.

Model organism association studies: *Genome-wide association studies in mammalian non-rodent models for the identification of genes relevant to human health and disease.*

LUPA, Unravelling the molecular basis of common complex human disorders using the dog as a model system. Despite major efforts, identifying susceptibility genes for common human diseases – cancer, cardiovascular, inflammatory and neurological disorders – is difficult due to the complexity of the underlying causes. The dog population is composed of ~ 400 purebred breeds; each one is a genetic isolate with unique characteristics resulting from persistent selection for desired attributes or from genetic drift/inbreeding. Dogs tend to suffer from the same range of diseases than human but the genetic complexity of these diseases within a breed is reduced as a consequence of the genetic drift and – due to long-range linkage disequilibrium – the number of SNP markers needed to perform whole genome scans is divided by at least ten. Here, they propose a European effort gathering experts in genomics to take advantage of this extraordinary genetic model. Veterinary clinics from 12 European countries will collect DNA samples from large cohorts of dogs suffering from a range of thoroughly defined diseases of relevance to human health. Once these different cohorts will be built, DNA samples will be sent to a centralized, high-throughput SNP genotyping facility. The SNP genotypes will be stored in central database and made available to participating collaborating centres, who will analyze the data with the support of dedicated statistical genetics platforms. Following genome wide association and fine-mapping candidate genes will be followed up at the molecular level by expert animal and human genomics centres. This innovative approach using the dog model will ultimately provide insights into the pathogenesis of common human diseases – its primary goal.

Implications of New Projects

At the moment, these topics in the integrated programme proposed for genetic variation studies might be funded at the level of €10 million to €12 million each, spread over 4–5 years. If all the negotiations for potentially funded projects are successful, then:

- New tools would be available to rapidly generate variation data.
- Epidemiological studies would be able to correlate genetic variation with disease characteristics.
- Genetic variation of microbes would lead to better understanding of disease processes.
- Model organisms suited to disease identification would contribute to the knowledge base.
- The unified databases and tools would be available as repositories and provide highly improved capabilities of linking and analysing data.

Part IV
Science Management, Perspectives and Conclusions

Chapter 10
Science Management

Abstract This chapter is devoted to aspects of science and project management. Involving much more than linking computer tools and databases, collaborative research depends on people, their needs and motivations, the institutional and legal and management frameworks. Otherwise collaborative research degenerates into a funding envelope with individual researchers following separate lines of research. The way in which the European Commission Framework Programmes address these challenges is described in what follows. Drawing on officially presented material, summaries are presented of how to participate and develop proposals, evaluation of projects and how to negotiate successful proposals. Extracts from the Seventh Framework Programme negotiating guidelines are presented, along with operating project attributes and management. Contractual and financial obligations are essential in motivating collaboration. Information dissemination, including the role of intellectual property rights and exploitation, is a key outcome of projects. Local and worldwide scientific collaboration methods are discussed.

Introduction

Motivation

This chapter is devoted to the aspects of project management. Involving much more than linking computer tools and databases, collaborative research depends on people, their needs and motivation for collaboration, the institutional and legal and management frameworks. Otherwise collaborative research degenerates into a funding envelope with individual researchers following separate lines of research. The way in which the European Commission Framework Programmes address these challenges is described in what follows.

F.B. Marcus, *Bioinformatics and Systems Biology,*
DOI 10.1007/978-3-540-78353-4, © Springer-Verlag Berlin Heidelberg 2008

Framework Programmes for Research

The European Commission supports health research, and many other areas, via the Framework Programmes for Research, described at the FP5 (2007), FP6 (2007) and FP7 (2007) websites of:

- CORDIS (2007), the European Community Research and Development Information Service
- Europa (2007), the Gateway to the European Union

Research in bioinformatics and systems biology was primarily supervised in the Health Research Directorate (Health-Research 2007) by the unit for fundamental genomics (Fundamental-Genomics 2007) in FP6 (2007) and currently by the unit for genomics and systems biology (Genomics-Systems-Biology 2007) in the Seventh Framework Programme (FP7), There are also major activities elsewhere in the European Commission research schemes, including Research Infrastructures, Directorate General for the Information Society and the Innovative Medicines Initiative. Full information about the entire Framework Programme is available on the FP7 (2007) website, with supporting documents available at FP7-Find-Document (2007).

Bioinformatics and Systems Biology Research

Within the FP7-Specific-Programme (2007) which will last from 2007 to 2013, an overall European strategy for research is defined. In health research, the areas related to genomics and systems biology are described as follows:

- High-throughput research: To catalyse progress in developing new research tools for modern biology including fundamental genomics that will enhance significantly data generation and improve data and specimen (biobanks) standardisation, acquisition and analysis. The focus will be on new technologies for sequencing; gene expression, genotyping and phenotyping; structural and functional genomics; bioinformatics and systems biology; other "omics".
- Large-scale data gathering: To use high-throughput technologies to generate data for elucidating the function of genes and gene products and their interactions in complex networks in important biological processes. The focus will be on genomics; proteomics, "RNA-omics"; population genetics; comparative, structural and functional genomics.
- Systems biology: The focus will be on multidisciplinary research that will integrate a wide variety of biological data and will develop and apply system approaches to understand and model biological processes in all relevant organisms and at all levels of organisation.
- Human development and ageing: Use of a wide variety of methods and tools to better understand the process of life-long development and healthy ageing. The focus will be on the study of human and model systems, including interactions with factors such as environment, genetics, behaviour and gender.

Methods for Executing the Specific Programme

In order to carry out the specific programme, a number of actions are taken to specify more precise, but still general research areas for projects, decided upon by a process of consultation, evaluation and assessment with the involvement of the research community and other players at all levels:

- Identification of research areas: A set of research areas is chosen via workshops, consultations, expressions of interest, contact with the scientific community, and after discussion with several supervisory committees. The research topics may be relatively narrow to strategically enable a particular field, or rather broad to allow the maximum initiative from the scientific community.
- Call for proposals: These topics are published in the *Official Journal of the European Union* (Official-Journal 2007), with the reference available on the CORDIS (2007) website, along with specific rules published on how to apply for funding in these areas.
- Preparation of proposals: Teams of researchers organise themselves into groups and prepare proposals for evaluation in response to these published topics involving collaborative research, following the rules set down in the so-called work programmes, which describe the topics and their overall research context and related documents. There is enough scope in the topics for research teams to specify their own preferred state-of-the-art approaches.
- Selection of projects: The eligible proposals received before the deadline are sent to reviewers and to evaluators, who after a first individual evaluation of the proposals come to Brussels to compare and reconcile their individual reviews of each proposal. There are up to ten reviewers involved per large proposal, in order to reach the best possible and most appropriate consensus on the evaluation, commentary, scoring and ranking of each proposal.
- Contract or grant agreement negotiation: Once the ranking lists have been approved and available funding has been provisionally assigned, a European Commission scientific officer negotiates a contract or grant agreement with the proposers, on the basis of analysis of the proposal and the evaluators' comments. These FP6 (2007) contracts or FP7-Model-Grant (2007) grant agreements contain general terms and conditions by which money is given to researchers to carry out the work described in their proposals.
- Operation of projects: Once a contract/grant agreement has been concluded and the project has begun operation, the scientific officer monitors progress by reading annual reports (overall summaries) and deliverables (generally detailed reports of specific activities) and reviewing milestones (dates for completion of activities), and holding periodic review meetings with the advice of external experts and scientific advisory boards for each project, where appropriate.
- Achievements: The achievements in bioinformatics and systems biology research are often made publicly available via project websites, as well as by normal publication channels.

Funding Methods

The European Commission funded several general types of projects in FP6-instruments (2007), which allow different combinations of researchers, depending on the scale of the task:

- Networks of excellence and integrated projects, typically €8 million to €15 million of Commission funding, spread over 4–5 years, involving 15–25 laboratories, concentrating on a broad research area
- Specific targeted research projects, typically €1.5 million to €3 million of Commission funding, spread over 3 years, involving five to 12 laboratories, concentrating on a focused research area
- Coordination actions and specific support actions, typically €1 million or less, for coordination or for particular targeted actions like a series of workshops. These types of activities do not directly fund research

In FP7 (2007), the integrated projects and specific targeted research projects are essentially combined and replaced by collaborative research projects of a similar nature, which may be of different-sized funding scales spanning a similar range to the FP6 (2007) funding methods and levels. Networks of excellence have been modified (FP7-Participate 2007).

Topics in Calls for Proposals

Topics at the beginning of FP6 (2007) were chosen by a fully open expressions-of-interest exercise (FP6-EoI 2007), which allowed all scientists to send in their concise suggestions for scientific areas of interest requiring a large-project approach. These suggestions were reviewed by panels of external experts, and the best areas were chosen as the basis for one or more topics in subsequent open calls for proposals. In FP7, the possibility of a "two-stage" procedure is being discussed, with a similarity to the expressions-of-interest exercise in the aspect of a short proposal, where, for example, proposers would send in brief proposals, but for a very specific approach, to a first stage of evaluation. For those short proposals successfully evaluated, and unlike in the expressions-of-interest exercise, the same team would be invited to submit full proposals in that precise area to a second stage of evaluation.

Proposal Evaluation

Fundable collaborative research projects are always chosen by teams of independent evaluators. Typically five external reviewers plus five evaluators who come to Brussels for evaluation contribute to the review of each large-scale proposal, which the evaluators themselves consider as a very detailed, careful, fair and open process. The evaluators are top scientists from Europe and around the world, who formally declare according to several formal criteria that they do not have any conflict of interest in the proposals they are evaluating. These rules and procedures are available on the FP6-evaluation (2007) website. Many evaluators who participate are very impressed with the high quality, and often in comparison with national procedures. After evaluation of individual proposals, the proposals are put in ranked order, and the top-ranked proposals are funded in ranked order up to the total amount of funding available. A contract is then negotiated with the Commission, based on the proposal and the evaluators' suggestions for areas that could be improved. For an overview of the whole process of applying to the call for proposals, see FP6-step-by-step (2007).

Nature of Contracts and Grant Agreements

The nature of European Commission research contracts is often misunderstood. The contract in general codifies what the researchers themselves propose, in the context of model contract general terms and conditions. In FP6 (2007), research consortia are highly self-governing, with internal control over fund distribution. They propose an updated work programme every year, conforming to general project goals. The project organises itself into work package teams which involve different clusters of project members, each with deliverables and milestones as goals.

The advantages of the work package and deliverable structure are:

- The Commission is able to monitor scientific progress and verify spending.
- Good internal project communication is provided.
- Public deliverables provide a detailed insight into collaborative research.

Project Management

Internal project management is also central to the success of the projects. A project involves much more than dividing money among participants who then carry out their individual research. Projects are instead highly interactive, with scientific decisions and actions decided at appropriate levels, including project coordinator, project management and scientific boards, annual meetings of the whole project, work package members, work package and deliverable coordinators, laboratory teams and individual scientists. See, for example, the management structure of BioSapiens-Management (2007). It all works very well usually, and when there are problems, there are also means to resolve conflicts and move forward and, if needed, to change scientific direction while staying within the general aims of the project.

In what follows, we discuss these areas in much more detail in the areas of:

- How to participate and develop proposals
- Evaluation of projects and how to negotiate successful proposals

- Extracts from FP7 negotiating guidelines
- Operating project attributes and its management
- Contractual and financial obligations
- Information dissemination
- Intellectual property rights (IPR) and exploitation
- Local and worldwide scientific collaboration

How To Participate and Develop Proposals

How To Participate in Research Projects

Guidelines for participation are given in detail in FP7-Participate (2007). While the details for participation in FP7 (2007) are fully available in the "Rules for participation", this section provides a summary of some of the main aspects. Further details can be found in the European Commission's proposals for the Rules for Participation (EC Treaty and Euratom), as follows:

1. Who can participate?
   - General provisions.
   - Eligible countries.
   - Eligible consortia.
   - Appointment of independent experts.
2. When are calls for proposals issued?
   - Calls (for proposals, for experts, for services and competitive calls).
   - Adoption roadmap.
3. What money and themes are available for funding?
   - Budget.
   - Funding schemes.
   - Research themes.
4. How is money given out to participants?
   - Financial rules/forms of grants.
   - IPR.
   - Consortium agreement.

European and International Participation and Collaboration

Researchers from the European Union member states and associated countries are the main participants in European collaborative research projects, but formal participation from almost anywhere in the world is possible and is encouraged. In addition, many projects have informal collaborations with research organisations and individuals around the world. As stated in FP7-Participate (2007), any company, university, research centre, organisation or individual, legally established in any country, may participate in a collaborative project provided that the minimum conditions have been met in the "Rules for participation" (FP7-Rules 2007). While FP7 participants can in principle be based anywhere, there are different categories of country which may have varying eligibility for different specific and work programmes:

- Member states: The 27 countries of the European Union;
- Associated countries: With science and technology cooperation agreements that involved contributing to the Framework Programme budget.
- Candidate countries: Currently recognised as candidates for future accession to the EU.
- Third countries: The participation of organisations or individuals established in countries that are not European Union member states, candidate countries or associated countries should also be justified in terms of the enhanced contribution to the objectives of FP7.

Calls for Proposals

The Commission periodically issues calls for proposals, and scientists can participate in the competition by referring to call texts, which are updated in the health area once or twice a year at FP7-Find-a-call (2007). All the information required is then available for each call topic. For example, see the closed call named FP7-CALL-HEALTH-2007-A (2007), which opened on 22 December 2006 and closed on 19 April 2007, with an electronic proposal submission deadline of 17:00:00. It is unwise to wait until the last hour to submit proposals, as many people do, since internationally linked computer systems are fallible, whereas the deadline is fixed.

All of the details related to the proposal process are officially available, and it is not hard to navigate the CORDIS (2007) website with a bit of experience and guidance, which hopefully this book is providing. The amount of information available is massive and detailed, some of which is copied below from the call as an example:

1. FP7-HEALTH-2007-A
2. Identifier: FP7-HEALTH-2007-A
3. Publication date: 22 December 2006
4. Budget: €637,000,000
5. Deadline: 19 April 2007 at 17:00:00 (Brussels local time)
6. Official Journal of the EU (OJ) reference: OJ C316 of 22 December 2006
7. Specific Programme: (Cooperation)
8. Theme: (Health)
9. Restrictions to participation: See eligibility criteria in the work programme
10. Information package: In order to receive a complete information package for this call, you will need to select the following elements

- The call fiche
- The work programme
- FP7 fact sheets – an overview of the basic features of this programme
- The guides for applicants relevant to the funding schemes used in this call
  - Call fiche
  - Work programme – general introduction
  - Work programme – health
  - Work programme – general annexes
  - FP7 fact sheets
  - Guide for applicants (collaborative projects – CP)
  - Guide for applicants(coordination and support action: coordinating – CSACA)
  - Guide for applicants (coordination and support action: supporting – CSASA)

11. Additional documents
    - European Parliament and the Council decision of 18 December 2006 concerning the FP7 EC (2007–2013)
    - Regulation laying down the rules for the participation to FP7 EC (2007–2013)
    - Council decision concerning the Specific Programmes
    - Rules for submission of proposals and the related evaluation, selection and award procedures
    - Evaluation forms

In particular, the "Guide for applicants" gives very clear instructions and templates as to how to prepare a proposal.

Mechanics of Proposal Preparation

Proposal preparation is carried out by electronic submission, with fixed templates shown in FP7-Proposal-Preparation (2007), via the Electronic Proposal Submission Service (EPSS 2007), which is an Internet-based application providing a secure work space for a consortium to prepare and submit a proposal jointly. Access requires only a standard Web browser; no special software has to be installed on the users' computers. For all details see the EPSS (2007) user guide. There are also a wide range of support services available for potential applicants at FP7-Get-Support (2007).

Support Services for the Mechanics of Proposal Preparation

The network of national contact points (NCPs 2007) is the main structure for providing local guidance, practical information and assistance on all aspects of participation in FP7. NCPs are national structures established and financed by governments of the 27 European Union member states plus the states associated to the framework programme. NCPs give personalised support on the spot and in proposers' own languages. The NCP systems in the different countries show a wide variety of architectures, from highly centralised to decentralised networks, and a number of very different groups, from ministries to universities, research centres and special agencies to private consulting companies. This reflects the different national traditions, working methods, research landscapes and funding schemes. Other support services include the FP7 support services. There is an enquiry service (FP7-Enquiry 2007) (a service provided by the Europe Direct Contact Centre) where it is possible to ask questions about any aspect of European research in general and the European Union Research Framework Programmes in particular. IGLO (2007) is an informal association of Brussels-based non-profit research and development liaison offices. The aim of IGLO is to facilitate and enhance the interaction, information exchange and co-operation between members of IGLO, their national research systems and the European institutions on issues related to European Union research and development, in particular, the Framework Programmes.

Finding Partners for Proposals

There is also a service to help find partners (FP7-Partners 2007). Building international partnerships is part of participating in European Union research programmes. CORDIS (2007) has an established partners service and a specialised service for FP7, fostering public–private partnerships to design, propose and launch new projects. You can use the search facilities to find international partners with complementary expertise, profile or technology. In addition, researchers usually know the top people in Europe in their field, and these people will very often be in the process of assembling teams for responding to the call for proposals.

Documents Relevant to the Proposal Process

There is a repository of relevant documents at FP7-Find-Document (2007):

1. FP7 legal basis
   - FP7 EC
   - FP7 Euratom
   - Specific Programme Cooperation
   - Specific Programme Ideas
   - Specific Programme People
   - Specific Programme Capacities
   - Specific Programme Nuclear Research
   - EC rules for participation
   - Euratom rules for participation

2. Legal documents for implementation
   - Rules for proposal submission
   - European Research Council (ERC) rules for proposal submission
   - Standard model grant agreement
   - ERC model grant agreement
   - Marie Curie model grant agreement
   - Rules on verification of existence, legal status, operational and financial capacity
3. Guidance documents
   - Guidance notes on audit certification
   - Guide for beneficiaries
   - Guide to financial issues
   - Guide to IPR
   - Checklist for the consortium agreement
   - Negotiation guidance notes
   - Reporting guide
4. Ethics review
   - Ethics check list
   - Supporting documents

Characteristics of Proposals

"Good" proposals are here considered as those that receive high scores for the evaluation criteria described later in this chapter and in "Rules for submission, evaluation, selection, award" at FP7-Find-Document (2007). However, each evaluator brings his/her own fresh interpretation to evaluations, within the official guidelines he/she is given.

An example of a proposal which became a funded project is EMBRACE (2007). Annex I of the contract with the European Commission, which is similar to the format and content of the original proposal, although altered by the negotiation process, also see below, has been posted on its website as EMBRACE-Proposal (2007).

In commenting and expanding on official material at CORDIS (2007), it seems that there are characteristics of "good" proposals, as well as misconceptions about the proposal submission and evaluation process. Listed in a similar order to the evaluation criteria, "good" proposals sometimes include:

1. Topic
   - Targeted orientation: The proposal corresponds to the printed call text. Proposals from a different area which are tidied up and resubmitted under a different area are common, but receive very mixed responses from evaluators.

- Financial instrument: With topics, specific financial instruments and indicated or maximum allowed budgets are specified. Proposals must respect specified limits, and work within them to obtain the best and most competitive proposals. The projects also balance the team size, ambition, management structure and other aspects against the actual financial limit and scientific scope of the project.
- Broad versus narrow topics: Topics that are very well described serve as a direct basis for well-focused proposals. Broad topics allow much more discretion to the proposers.
- Competition and context of other topics: Good proposals are shaped by the entire call text, which provides the context of a topic, the limitations and the range of competition from other proposals. These topics and their context also guide the scientific interests of the people who will be chosen to be independent evaluators of these proposals, whose names are available from an evaluator database.

2. Science

   - Top quality: Good large-scale proposals have received up to ten reviews by independent evaluators, who are often the best in their scientific field, and who are legally bound to be without a conflict of interest. Therefore, only top quality proposals will succeed. Typically there are one to three high-quality proposals per narrowly focused topic, so the success rate for a good proposal is high.
   - Wet and dry laboratory capability: In the area of bioinformatics and systems biology, and especially in systems biology, the most successful projects have often had either strong links to or direct participation of wet-laboratory experimentalists. This makes for better science and a better proposal.
   - Excellent explanations: These proposals are well thought out and well described. Especially in the introduction, the project plan is carefully explained, including motivation, context, management and financial plans, scientific background, goals, motivation and hoped-for results. It is often optimistically assumed that evaluators will have a shared vision with the proposers. In fact, it is necessary to describe the overall vision clearly and in detail, starting from basics.
   - Scientific approach: The scientific problems are clearly described, and solutions are proposed to major scientific problems. Often, strategically important new resources are proposed. The methods and overall strategy are clearly described, including the experimental and analysis chains of events and how they all interrelate.
   - Previous work and attempts: It is often the case that innovative proposals involve work that has been tried and has failed elsewhere. This history should be clearly described and confronted, and it should be shown how the new approaches will provide a successful solution.
   - Current state of the art and synergies: Often there are either single or collaborative projects already existing in the area of the project or related areas. Good proposals take full account of these, describe their status and often

develop valuable synergies with these other projects, while at the same time establishing a separate identity.
- Technological development: When technological development is involved, and depending on the topic, it is often important to include a scientific research component which either uses or demonstrates how a technology will be put into practice. Tool development without proven application can often lead to a result that has little outside interest.
- Limitations: There are always practical and scientific limitations to what can be accomplished owing to nature itself, and by a project that is finite in resources and time. These limitations should be clearly described, and it should be shown how the project is addressing and dealing with these limitations.
- Data: All aspects of data handling need to be addressed, including sources, ethical aspects, data accuracy, data curation and consistency.
- Surveys: When the wider scientific community is involved, surveys at the start of the project are very useful for ensuring uptake and usage of any new resources and results. This is not a substitute for having a clear major option, which may be altered according to the results of the survey.
- Ethics and other Commission policies: Special care should be taken to be sure that projects confirm to the specified ethical guidelines. Local ethical committee procedures should be cited. Other Commission policies as discussed in the guidelines such as those relating to gender, etc. should be properly addressed.

3. People and Management

   - Management plans: The management plans, budgets and person-months required should be well thought out and extensively defended. The ambition of goals should be clearly related to the budget in money and manpower proposed.
   - Management structure: Good existing projects show the sorts of organisations that work well. These usually include the co-ordinator, a project manager, a financial officer, an executive board, work package and activity managers, integration meetings and outside review boards and exploitation boards.
   - Capabilities: It is important to describe the full range of capabilities of people and laboratory resources that can be brought to bear on the project. The detailed plans need to be clearly described.
   - Teams: Good proposals have often resulted in the top people in Europe in the field being assembled, while keeping the overall number of teams at a manageable size, with each partner chosen for what it can specially bring to the consortium. These can often include partners from an institution that has not participated in collaborative research before, but where a group has developed a specialty that makes it an ideal partner in a particular area. Such groups then go on to take a more major role in future collaborations, thus providing an entry for smaller and less experienced partners.
   - Small and medium-sized enterprises (SMEs): These are often included in projects for their scientific or commercial expertise, which is often not present in a pure research environment, and for their special abilities in exploitation

and commercialisation of results, where appropriate. Indeed, one of the major justifications of the existence of the Framework Programmes in the original Treaty of Rome is the requirement to increase European competitiveness.
- Integration: The work packages and teams need to be well integrated and interactive. A series of non-communicating individual projects bundled together to get funding does not bring added value from the collaborative process.
- Standards and quality control: The whole concept of quality control should be imbedded in a project. There should be continuous evaluation at all levels. Data, tool development, experiments, standard operating procedures and results are obvious areas, but broadness of goals against resources, quality of work, scope of the work and relevance to project main goals should be the subject of continuous assessment, with structures provided to make this happen.
- Skills transfer: One expert partner does not ensure an excellent collaboration. Structures need to be established to ensure within-collaboration training, exchange of people, skills, materials and capabilities.

4. Expected Results and Impact
   - Knowledge management: Proposals should have clear and extensive plans for results, publications, deliverables and dissemination, especially via a project public website established at the beginning of a project. In particular, top journals should be targeted. There should be an important emphasis on training, teaching and workshops.
   - Exploitation : When exploitation is in the public domain, clear plans are needed as to how access will be provided, and what are the benefits to users. If commercial aspects are involved, there should be a full survey of pre-existing IPR inside and outside the consortium, and plans for future IPR. The state of the art needs to be clearly described. If treatment of disease aspects is involved, the plan should be especially clearly described. Exploitation will often depend on the degree of integration of work, data handling and bioinformatics within the project.
5. Misconceptions
   - Hidden agendas: People want to know the hidden agenda behind a topic. There is none. It is true that topics and research areas are chosen by an extremely extensive consultation process, and finally approved by several levels in the Commission, plus scientific advisory committees plus input from the member states' representatives. However, once the text has been published in the *Official Journal of the European Union* and on the CORDIS (2007) website, the text is all there is. The external evaluators compare the proposals with the printed text detailing the call topic, without any reference to the process of choosing the topic. There cannot be a hidden agenda.
   - Is a topic in scope? The most common question that people ask the Commission staff is: "Is my topic in scope?" This is the wrong question, and is related to the concept of hidden agendas. Once a topic has been published, unless a proposal is wildly and obviously out of scope, the only people who decide whether or not

the proposal is relevant to the published topic are the independent evaluators, and all they refer to is the printed topic itself and its context of general research areas in the work programme. Hence proposers can get the best answer to the question by asking either themselves or a colleague who might someday be an evaluator for their opinion. A better way to put the question is: "Do I think that my proposal is relevant to the published topic, within its context in the work programme, and do I think my proposal is really excellent and will be chosen above competing proposals by evaluators who are scientists just like myself?"

- Lobbying: People think that lobbying can influence a choice or rejection of a proposal after the evaluation. It cannot!
- Budgets: It is thought that proposed project budgets should always be higher than appropriate, since budgets will always be reduced during the evaluation process. Not true. A well-defended proposal is often fully funded, depending on overall Commission funding available.
- Topical funding: Not all topics will be funded. Only excellent proposals will be funded, and if none are good enough, proposals for those topic will not receive funding.
- Institutional size: Some people feel only the largest institutions can participate. Although the larger institutions do have large administrative departments and experience of proposal preparation, good proposals often come from smaller institutions, universities or SMEs, for the first time, who can also be the coordinators. High scientific quality, enthusiasm and attention to detail can result in good proposals. For first attempts, smaller institutions usually join with larger institutions to gain experience.
- The biggest misconception is that the European Commission only produces inappropriate legislation (Euromyths 2007) and makes life difficult for researchers. Hopefully this book has demonstrated the important and positive involvement of the Commission in research and other areas besides. Much of the scientific administration in projects is necessary to produce a true collaborative atmosphere, with major benefits for all. Much of the financial administration is devoted to justifying and thereby allowing very advantageous cash advances from the European Commission to each of the project partners at the very beginning of a project, and ensuring the money is actually spent on project-related research during the project.

Evaluation of Proposals

Evaluation Process

In order to develop a proposal, it is vital to understand the evaluation process. Funded projects are *always* chosen by teams of independent evaluators. Proposers are not informed of the names of those who evaluate their proposals. For large-scale projects (€12 million), typically five external reviewers plus five evaluators who come to Brussels for evaluation contribute to the proposal review, which is a very detailed, careful, fair and open process. The evaluators are top scientists from Europe and around the world, who do not have any conflict of interest in the proposals they are evaluating. These rules and procedures are available on the FP6-evaluation (2007) website. Evaluators who participate are often impressed with the high quality of the process, often favourably in comparison with national procedures.

Evaluation Criteria

The independent evaluators evaluate according to fixed evaluation criteria, see "Rules for submission, evaluation, selection, award" at FP7-Find-Document (2007). The detailed evaluation criteria, subcriteria and associated weights and thresholds are set out in the work programmes, based on the principles given in the Specific Programmes, and on the criteria given in the "Rules for participation". In the specific case of FP7-CALL-HEALTH-2007-A (2007), a number of key points are made. The general eligibility criteria are set out in Annex 2 to this work programme, as follows, for collaborative projects, and these three, equally weighted and equally important criteria, receive scores and comments which are reported on the evaluation summary report:

1. Criterion 1: Scientific and/or technological excellence (relevant to the topics addressed by the call)
   - Soundness of concept and quality of objectives
   - Progress beyond the state of the art
   - Quality and effectiveness of the S/T methodology and associated work plan
2. Criterion 2: Quality and efficiency of the implementation and the management
   - Appropriateness of the management structure and procedures
   - Quality and relevant experience of the individual participants
   - Quality of the consortium as a whole (including complementarity, balance)
   - Appropriateness of the allocation and justification of the resources to be committed (budget, staff, equipment)
3. Criterion 3: The potential impact through the development, dissemination and use of project results
   - Contribution, at the European (and/or international) level, to the expected impacts listed in the work programme under the relevant topic/activity
   - Appropriateness of measures for the dissemination and/or exploitation of project results, and management of intellectual property

In addition, there are general eligibility requirements for the particular call in question. It is important to note that the following funding thresholds will be applied as eligibility criteria and that the proposals which do not respect these limits will be considered as ineligible:

1. Small or medium-scale focused research projects: The requested EC contribution shall not exceed €3 million unless otherwise indicated in the topic description.
2. Large-scale integrating projects: The requested EC contribution shall be over €6 million and not exceed €12 million unless otherwise indicated in the topic description.
3. The minimum number of participating legal entities required, for all funding schemes, is set out in the "Rules for participation" and is presented in the relevant parts below. A collaborative project shall have at least three independent legal entities, each of which is established in a member state or associated country, and no two of which are established in the same member state or associated country.
4. A proposal will only be considered eligible if it meets all of the following conditions.

   – It is received by the Commission before the deadline given in the call text.
   – It involves at least the minimum number of participants given in the call text.
   – It is complete (i.e. both the requested administrative forms and the proposal description are present).
   – The content of the proposal relates to the topic(s) and funding scheme(s), including any special conditions, set out in those parts of the relevant work programme

Announcement of Evaluation Results

The evaluators prepare evaluation summary reports for each proposal summarising the results. All proposers after evaluation receive a copy of the evaluation summary report with comments and scores by the independent evaluators. If and when a proposal has passed all evaluation thresholds, and if the project is ranked highly enough compared with competing proposals to qualify for the limited funding available, typically but not always one proposal per published topic, the European Commission first sends a letter opening negotiation, and then the project co-ordinator negotiates with the Commission scientific officer. The main items of the negotiation and major changes are discussed in the evaluation summary report.

For those proposals that pass all thresholds but are not funded, they are placed on a reserve list, with a low probability that money will become available to fund them. Because of this low but finite probability, the Commission may not communicate with the proposers until it is certain that money is not available. For those proposals below evaluation thresholds, the proposals will not be funded. In FP7, a redress procedure has been made available, but is available only for questions of procedure, e.g. a mistaken rejection for non-eligibility, not for disagreement with the scientific, managerial and impact assessments of the evaluators. It does not give the opportunity to revise or resubmit a proposal.

Project Negotiation

FP7 Negotiating Guidelines

In FP7-Negotiating (2007), a number of guidelines are presented. The following is a summary of some key features.

The What, Why and How of Negotiations

The overall purpose of negotiations is to finalise the details of the work to be carried out under the grant agreement within the associated budget, as well as to establish the legal and financial information needed to establish the grant agreement.

The project negotiation process comprises two main aspects:

1. Technical (*scientific*) negotiations
2. Financial and legal negotiations

Technical Negotiations

The aim of the technical negotiations is to agree on the final content of Annex I (description of work) to the European Commission Grant Agreement. Templates and instructions for preparing Annex I are available at CORDIS (2007). During this part of the negotiation process:

- The proposal may need to be adapted to meet the recommendations of the evaluation, as described in the negotiation mandate.
- The Commission will verify that the project objectives are "SMART" (specific, measurable, attainable, realistic, timely).
- The full work plan of the project will need to be defined in sufficient detail.
- The work to be carried out by each of the beneficiaries and any potential future expansion of the consortium will need to be defined in sufficient detail.
- Agreement will need to be reached on the list of deliverables and their content, timing and dissemination level.
- Agreement will need to be reached on the project milestones and their assessment criteria.

- An indicative time schedule needs to be established for the project reviews (if not predefined in the special conditions of the grant agreement) – which ideally should be synchronised with the reporting periods.

Financial and Legal Negotiations

Financial negotiations rely on the FP7-Financial (2007) guidelines and the FP7-Beneficiaries (2007) guide and focus mainly on reaching agreement on budgetary matters such as the budget for the full duration of the project and the budget breakdown for the different project periods, as well as issues related to subcontracting and third parties. They will also cover the establishment of the amount of the initial prefinancing, timing of project periods and reviews. Legal negotiations include the analysis and review of the legal status of each applicant and the final composition of the consortium, any special clauses required for the project, and other aspects such as the project start date. During this part of the negotiation process:

- The total costs, total eligible costs and maximum Community financial contribution will be determined. Special attention should be given to the method to calculate the personnel costs and indirect costs.
- A table of the estimated breakdown of budget and Community financial contribution per activity to be carried out by each of the beneficiaries will be established.
- The start date and the duration of the project are agreed upon.
- The need for the inclusion in the grant agreement of any special clauses will be established.
- Where applicable, a roadmap will be established for any planned competitive calls relating to the later addition of new project partners and the budget available for the consortium expansion will be agreed upon.
- The timing of the reporting periods will be established (normally every 12 months).
- Any subcontracting or third-party issues will be clarified.
- At this stage the Commission will also assess whether the proposed coordinator has the required management skills, capabilities and experience to carry out the coordinator's tasks.

Negotiation Relevant Documents

See also FP7-Find-Document (2007), especially the guidance documents:

- Guide for beneficiaries
- Guide to financial issues
- Guide to IPR
- Checklist for the consortium agreement
- Negotiation guidance notes
- Reporting guide

Negotiation Key Points

The official negotiating guidance notes provide clear directions as to procedures. In summarising them, several areas often come up in negotiations, which are useful for consideration, summarised in the following informal list, which obviously have similarities with the characteristics of good proposals:

1. Reporting and deliverables: If each work package has at least one R (report), P(public) deliverable per work package per year, then it is easier to monitor the progress of a project. Also, as much information as possible is put into the public domain. If a project is highly commercial, then public disclosure is less appropriate and necessarily restricted. So, for example, with a 4-year project with ten work packages, 40–80 deliverables might be expected. According to reporting guidelines, deliverables should be delivered, each one having its own cover and in some form of report.
2. Reporting period: This is the period that is in the contract, and relates to financial and scientific reporting. A 12-month reporting period is appropriate for large projects, and 12–18 months for smaller projects. Since reporting periods by definition involve financial and scientific reports together, it is additionally possible to have more frequent scientific reports and deliverables. A deliverable is primarily documentation and proof of work done. Milestones have deadlines as do deliverables, but they are not reports. They represent goals and management aids.
3. Wet and dry laboratory capability: In the area of bioinformatics and systems biology, and especially in systems biology, the most successful projects have either strong links or direct participation of wet-laboratory experimentalists.
4. Team interaction: Management structures and work package organisation should ensure that teams interact, rather than act as a collection of separate projects. The work packages and teams need to be well integrated and interactive.
5. Data: All aspects of data handling need to be addressed, including sources, ethical aspects, data accuracy, data curation and consistency, and procedures for interpreting the data.
6. Skills transfer: One expert partner does not ensure an excellent collaboration. Structures need to be established to ensure within-collaboration training, exchange of people, skills, materials and capabilities.
7. Knowledge management: Projects should have clear and extensive plans for dealing with results, publications, deliverables and dissemination. In particular, top journals should be targeted. There should be an important emphasis on training, teaching and workshops.
8. Exploitation: Where exploitation is in the public domain, clear plans are needed as to how access will be provided, and what the benefits to users will be. If commercial aspects are involved, there should be a full survey of pre-existing IPR inside and outside the consortium, and plans for future IPR. The state of the art needs to be clearly described. If treatment of disease aspects is involved, the plan should be especially clearly described. The success of exploitation will often depend on the degree of integration of work and data handling and bioinformatics

within the project. There are already fixed IPR rules as a base; see the model grant agreement and IPR guidelines at FP7-IPR (2007).

9. Evaluation summary report and scientific programme: The project should always implement evaluation summary report recommendations, although there is often flexibility, where appropriate, to achieve project goals.
10. Co-coordinator management: Some projects delegate aspects of day-to-day management to an institution other than the co-coordinator. There are certain responsibilities that must stay with the co-coordinator, as described in the FP7-Negotiating (2007) guidelines.
11. Project management structure: There should be a complete structure in place, especially for large projects (see, for example, BioSapiens-Management 2007 for an academic type project, or AGRON-OMICS-Management 2007, where technology transfer is planned).
12. Subcontracting: Subcontracting is difficult to implement properly, and often requires open calls for proposals with several competing bids.
13. Consortium agreement: An FP7 checklist for consortium agreements (FP7-Checklist 2007) is provided under FP7 documents. There is no legal requirement regarding the contents of the consortium agreement, and agreements can in principle be of minimal complexity, since many problems are already addressed in the model contract, its annexes and the technical Annex I. However, there are two essential needs. One is a mechanism for conflict resolution. If one partner behaves badly, somehow the consortium must provide for some means of resolving the problem. There should also be a transparent way of agreeing on how to internally redistribute money in case it becomes necessary, for example, to apply for a contract amendment. This can be in the consortium agreement or in the main technical annex.
14. Website: Projects should have a public website, preferably with their programme, publications list and public deliverables reports posted there.
15. Test activities: Some projects which are developing tools will have a "test problem" section. If these activities are made applicable to the most interesting science and represented as thematic work packages, it is beneficial to all concerned.
16. Links with other projects: This is always to be encouraged, but there are formal and informal links. Legally, only project participants control how the project runs, and they are fully responsible for the results promised in the contract. It is contractual obligations that define who is a partner. Within these limitations, external informal collaboration is very common and often highly beneficial.
17. Each institution will have its own accounting rules, and the calculation of indirect costs is important, since it determines how much the scientists get, and how much the central administration gets for overheads. Institutions should be aware that the same model must be applied for all contracts in FP7. Each institution should check it has a consistent policy. There are four models for calculating indirect costs (overheads), which needed to be applied to calculate the amount of European Commission funding to request for each partner in the proposal and in the grant agreement:

- REAL and SIMPLE indirect costs are mostly for large industry with sophisticated accounting systems. The REAL indirect costs are overheads calculated and defended in detail.
- SIMPLE indirect costs are where overheads are calculated at the legal entity level and not per person per project.
- FLAT_TRANS indirect costs are calculated as 60% of personnel plus other direct costs. Then the European Commission contribution is 75% of that (note that 75% of 160% equals 120% of direct costs, which is similar but not identical to the Sixth Framework Programme (FP6) Additional Cost (AC) cost model). Many institutions are eligible for this.
- FLAT_STD is as FLAT_TRANS, but with 20% instead of 60%, and everybody is in principle eligible for this.

18. Management costs can be refunded by the European Commission at 100%. In FP6, the management costs could only be maximum 7% of the total European Commission contribution. In FP7, this limit does not exist; however, the 7% limit did represent an appropriate level.

Operating Project Attributes and its Management

The Model Grant Agreement

Once negotiation has finished, an FP7 grant agreement (a contract in FP6) is signed between the European Commission and the coordinator, with the partners in the collaboration individually acceding to the contract, the Commission provides prefinancing of activities and the project activities begin. The most important attribute of FP6 and FP7 contracts is that the scientific programme is chosen and managed by the participating scientists themselves, and that there is a great deal of flexibility in the arrangements. The scientific programme was chosen by the scientists in their proposal, and then details are negotiated with the Commission scientific officer in line with evaluators' comments, usually maintaining most of the original project objectives, depending on the scale of budget cuts compared with proposal requests. Once operating, the projects are self-managing. For the longer 4–5-year projects, the scientific programme was revised every year in FP6, and project revision remains as an option for FP7 via the amendment process, in order to update the research programme as progress is made or problems are encountered.

Project Coordination and Management Structures

Successful projects have established structures that are complicated enough to manage the wide range of activities, yet simple enough to function efficiently and to take decisions. They also will have established robust conflict resolution mechanisms, so that inevitable problems can be dealt with and the project moved forward. Good existing projects show the sorts of organisations that work well. These usually include the co-ordinator, a project manager, a financial officer, an executive board, work package and activity managers, integration meetings and outside review boards and exploitation boards. In particular:

1. The management structure is very important for large projects. BioSapiens-Management (2007) has responsibility for establishing the network and achieving its objectives, including project reporting and accounting. The coordinator is supported by a project manager, whose role is to provide day-to-day support and implementation of all management and organisational tasks. A steering committee, chaired by the coordi-nator, oversees the work of the network and assists the coordinator in the management of BioSapiens. The steering committee meets at least quarterly. In addition to the steering committee, a work package coordinating committee and a training committee oversee the progress of the work packages and training activities respectively.
2. The current members of the steering committee are
   - The project coordinator (as chair)
   - The training coordinator
   - The outreach coordinator
   - The work package coordinator
   - The project manager
3. The scientific advisory board is a critical part of this network, and includes 12 senior scientists, each representing a different aspect of biology and bringing considerable experimental knowledge and expertise. Their role is to help the network plan its work, and to promote the interactions between the experimental and bioinformatics communities. Each work package also has its own co-ordinator, and is responsible for the activities of the team within the work package.
4. Although the structure seems complicated, it functions extremely well, leading to
   - Extremely efficient exchange of information and work within and between work packages.
   - High standards of documentation.
   - Input from all levels when strategy is considered.
   - Enough devolvement that each participant and each researcher feels they have an important role to play, and enough freedom that individuals can not only publish to further their careers, but achieve top publications owing to the high and unique quality of the work.
   - Efficient distribution of funds and efficient project control.
5. There are variations to the BioSapiens structure, for example with the structure of AGRON-OMICS-Management (2007). There is a user committee that is more oriented towards possible commercial exploitation of the results of the consortium. Much depends on the direction of scientific and commercial emphasis of the projects, and the nature of the partners.

Role of the Commission Scientific Officer

Although the project has extensive autonomy, the Commission scientific officer has extensive authority and responsibilities specified in the contract or grant agreement, in the area of technical and financial review, as follows: "At the end of each reporting period, the Commission shall evaluate project reports and deliverables required by the provisions of Annex I and disburse the corresponding payments" and "After reception of the reports the Commission may: a) approve the reports and deliverables, in whole or in part or make the approval subject to certain conditions. b) reject the reports and deliverables by giving an appropriate justification and, if appropriate, start the procedure for termination of the grant agreement in whole or in part." Termination is very rare, but partial report rejection and requirements for rewriting are more common. It is thus necessary for the scientific officer to be able to understand and evaluate the whole of the technical knowledge in the reports, and evaluate if the work done corresponds to the spending made, even when supported by external experts. The scientific officer can use this authority to enforce high standards in documentation in reports and deliverables, and help to encourage the project with constructive oversight, while leaving the technical management and decision completely with the project internal management and with the researchers. In addition to other responsibilities within the European Commission such as dealing with outside queries and responding to interservice consultations and internal administration requirements, a scientific officer has the multiple research-related responsibilities of:

- Developing consultations and thematic workshops involving leaders from several projects meeting together with outside experts, to develop strategic perspectives
- Participating in drafting work programmes with call topics
- Organising evaluation of proposals, choosing external experts and moderating the evaluation
- Negotiating contracts
- Reading, understanding and commenting on all scientific publications, reports and deliverables produced by the projects, and correlating results with financial reporting and participating in the management chain for approving financial payments
- Organising periodic project reviews including outside experts
- Management oversight in the case of problems and explaining procedures to the members of the projects
- Helping projects to organise reviews, contract amendments when partners change
- Preparing publications, publicity and information about Commission-funded research

With a portfolio for each scientific officer of typically 20 active projects involving Commission contributions of €100 million, supplemented by participants' formal contributions and indirect contributions from related laboratory programmes, a critical mass is achieved for supervision of a research field at the programme level as well in individual research areas.

Role of Commission Financial and Contract Officers

The financial and contract officers of the Commission are the people that make the process work. They help to prepare the contracts and monitor and check financial statements in the periodic reports from the consortium, working closely with the scientific officer to maintain a high quality of oversight. The financial officers and the scientific officer make a team that helps ensure the smooth running of a project and the smooth flow of the funding that makes it possible. The financial and contract officers also have a key role in processing the amendments that allow a project to adapt to changing circumstances, for example the change of institutions involving a key research team, or a major change to the scientific programme.

Contractual and Financial Obligations

Role of Contracts in Collaboration

Contractual obligations have an important role to play in participants' behaviour, their incentives to work together and their incentives to exchange knowledge. Science training and the culture of research, especially in universities, may encourage people to work on their own, and to guard their data until well after a journal publication is produced. The details of contracts are also often felt to be exclusively for administrators, and nothing to do with research itself, apart from providing money. In collaborative research, people have to commit themselves to particular research agendas, but in return they are only contracted to do what they themselves have offered to do, they have a great deal of flexibility in choosing how work may be done, and they have access to a much wider range of knowledge and resources, and are probably able to publish more easily and quickly, owing to higher-quality results in many cases.

Model Grant Agreements

The details of a grant agreement are specified in FP7-Model-Grant (2007), which are signed by both participants and the Commission. The key texts are in the core text and Annex II (general conditions). Annex I, the technical annex, is very similar in form to the proposal, and contains the details of the technical work to be done, in particular the work programmes, deliverables and milestones. The FP7 Grant Agreement – Annex II – governs the way that projects function internally, and shows the role of the Commission in supervising them. The financial clauses are vital also. For those who want more information on financial matters, they can refer to the other parts of the model grant, and to the very detailed FP7-Financial (2007) guidelines. As might be imagined, there is a certain amount of administrative overhead in this administration, but the working scientists are partly shielded, especially as the Commission typically funds up to 7% of project budgets for management costs, generally allowing the project to engage administrative officers to handle these aspects.

Project Reporting

There are fairly strict guidelines on reporting and deliverables which are essential to the functioning of the project. As is evident, there is a formal reporting structure, but in practice it means that once a year or every 18 months each researcher and work package group needs to write up their reports for the project, for the Commission and for the public, where appropriate. The Commission, external reviewers and review boards give advice on progress and on future scientific direction. Among the participants in the projects discussed in this book, almost everybody feels that all concerned benefit from this reporting and review process, and that it strengthens and improves the scientific cohesion and output of the project.

Reporting tends to accomplish three main goals:

1. The Commission scientific officer can monitor the scientific progress properly and can judge if the work done corresponds to the money claimed on the project.
2. Reporting, including deliverables and accompanied by milestones, is a strong internal management and communication tool within the project. Regular reporting means that everybody inside the project knows what everybody else is doing, and in detail. This supplements the information exchanges within laboratories, within work package working groups and within project meetings. Web postings accelerate this process.
3. Reporting provides the outside world with supplementary, vital and up-to-date information on what is going on in a project, especially when the deliverables are public reports posted on the project website. The reports supplement published material and give much more detail about how research was done.

Information Dissemination

Publications and Websites

Each project determines its own style of information dissemination. All projects produce important journal publications, and also a number of deliverable reports that are more or less public, depending on the project. These are usually available via the project website.

A Systems Biology Website

One of the best documented is the ENFIN (2007) project. From the main website ENFIN (2007), there is direct access, via the publications tab, to the press releases, journal publications and publicly available deliverable reports, which constitute in themselves a major information source about the scientific and technical details of project results. The advantage of this unified format is that researchers have immediate access to the formal publications, to the tools that are related to them and, via the deliverables, to the research methods and team members that actually produced the results. Thus, a window is immediately furnished into all aspects of the research and resources generated. This is especially important in a project like ENFIN, one of whose goals is to develop toolboxes for the experimental community. There are Wiki's for internal communication also.

A Bioinformatics Website

In another project, BioSapiens (2007), publications are available on the website home page, and additional references and deliverables are available via each work package. There is also a major emphasis on dissemination via conferences, training and especially the outreach programmes and the European School of Bioinformatics. The trainings and meetings tabs of the BioSapiens website show the diverse and extensive activities of this project. Large projects have a major presence at scientific meetings, and can organise major workshops centred around their own activities.

IPR and Exploitation

IPR Rules

The European Commission encourages information dissemination and exploitation in its contracts, implemented via the rules for participation and dissemination (Regulation 2006). A European Commission FP7 grant agreement includes Annex II of the model grant agreement, which contains the IPR rules. Selected key rules are copied below to illustrate that the rules are short, but comprehensive enough to provide a full legal and operational background for projects. These rules are designed to be flexible enough to allow a range of project operation from fully open source and open publication approaches appropriate to basic research, or to allow and encourage full commercial exploitation of an intellectual property developed during the project:

1. II.26. Ownership.
   - *Foreground* shall be the property of the *beneficiary* carrying out the work generating that *foreground*. ("Foreground" means results, including information, whether or not they can be protected, which are generated under the project.)

- Where several *beneficiaries* have jointly carried out work generating *foreground* and where their respective share of the work cannot be ascertained, they shall have joint ownership of such *foreground.* They shall establish an agreement regarding the allocation and terms of exercising that joint ownership.
- However, where no joint ownership agreement has yet been concluded, each of the joint owners shall be entitled to grant non-exclusive licences to third parties, without any right to sublicence, subject to the following conditions
  - At least 45 days prior notice must be given to the other joint owner(s).
  - Fair and reasonable compensation must be provided to the other joint owner(s).
- If employees or other personnel working for a *beneficiary* are entitled to claim rights to *foreground*, the *beneficiary* shall ensure that it is possible to exercise those rights in a manner compatible with its obligations under this *grant agreement.*

2. II.28. Protection.
   - Where *foreground* is capable of industrial or commercial application, its owner shall provide for its adequate and effective protection, having due regard to its legitimate interests and the legitimate interests, particularly the commercial interests, of the other *beneficiaries.*
3. II.29. Use.
   - The *beneficiaries* shall *use* the *foreground* which they own or ensure that it is used.
   - The *beneficiaries* shall report on the expected *use* to be made of *foreground* in the plan for the *use* and *dissemination* of *foreground.* The information must be sufficiently detailed to permit the *Commission* to carry out any related audit.
4. II.30. Dissemination.
   - Each *beneficiary* shall ensure that the *foreground* of which it has ownership is disseminated as swiftly as possible. If it fails to do so, the *Commission* may disseminate that *foreground.*
   - *Dissemination* activities shall be compatible with the protection of intellectual property rights, confidentiality obligations and the legitimate interests of the owner(s) of the *foreground.*

Approaches to IPR Policy

The section on IPR in the model grant agreements was shortened from 30 pages in the Fifth Framework Programme to three pages in FP6 and was only slightly modified from FP6 to FP7. The modifications were partly informed by several workshop reports on the role of IPR in research, with two of the most relevant being Granstrand (2007) and Crespi (2007). Recommendations from the IPR-Bioinformatics (2007) report include the following IPR guidelines to research organisations:

- It is difficult for universities and SMEs to generate a significant revenue stream on IPR-protected databases and software, unless the databases are large and comprehensive. IPR protection should be used, where appropriate, to achieve broader strategic objectives such as encouraging collaborative project funding and building core knowledge, rather than just licensing income.
- Awareness training in the strategic use of all types of IPR for bioinformatics is needed for researchers and technology-transfer offices.
- Universities and public research institutes need to review the terms on which they manage IPR generated by the work of their employees, especially where such work contributes to infrastructure (such as databases and related software tools) of relatively permanent value, but whose value and maintenance depend upon the continuing personal commitment of individuals who may move or lose interest. A consistent policy on ownership of database rights, responsibilities, benefits sharing and support should be formulated by public research bodies.
- International collaborative research needs to take account of different IPR legal systems in the European Union and the USA, especially concerning the grace period (publishing before patenting), and the laws relating to databases, copyright and software.
- IPR ownership and the roles of electronic database publishing companies pose new challenges that need careful and well-understood rules and best practices for all parties concerned, with a balance of rights and responsibilities.
- Concerning IPR-controlled access to databases and software, a period of experimentation may be appropriate, ultimately leading to better co-ordinated policies addressing various goals, while not compromising on the need for publicly available bioinformatics databases without charges.

Open Source

The discussions and recommendations in these reports reflect the tension between those who feel that all Internet-based research should be fully open source and fully accessible without charge, and those who see that the data may have a high commercial value. The tension is often between the user, who wants full access, especially in the age of linked Web services, and the database producer and owner, who often struggles with poorly adapted and short-term infrastructure funding policies at various levels. Many databases therefore are not properly supported, leading the owner to seek revenue streams to maintain the database. In the commercial world, data and software are seen as both valuable in themselves, and also key to commercial advantages that a company might have. In the projects discussed in this book, many participants have the philosophy of open source access to many of the databases and some of the tools, but others maintain IPR protection over the software developed, and others maintain IPR protection over results where important drug development or other commercial exploitation is possible. Again, it depends on the choice and nature of the participants and the goals of the proposal.

IPR and Commercial Exploitation Support Services

The European Commission provides a wide range of support services, helpdesks and organisations to advise on IPR and on commercial exploitation of results, available through the CORDIS website (FP7-Other-Support 2007) as non FP7 (2007) specific support services. Other support services provide additional help and guidance on European Commission funded research and innovation related issues:

- IPR HelpDesk : To assist potential and current contractors taking part in Community-funded research and development projects with IPR issues. The helpdesk offers two main services: an informative website open to all interested parties, and a free legal helpline aimed at participants in European Commission funded research under the Framework Programmes.
- Gate2Growth Initiative: A unique portal, bringing together a community of entrepreneurs, investors, service providers and several networks supported by the European Commission that provides information and guidance on innovation financing sources and helps to locate professional expertise in innovation financing issues.
- Innovation Relay Centres (IRCs): These are a network of centres in the European Union and beyond providing local help to promote technology partnerships and transfer. The goal of the IRC network is to promote innovation, to encourage exchange of research results between organisations across Europe and to provide advice, consulting and training support which meets the specific needs of each company and its local industrial situation.
- Euro Info Centres (EICs): As an interface between European institutions and local companies, the EICs provide information, advice and assistance to SMEs in all Community matters.
- Business Innovation Centres: Regional structure of support to innovative SMEs and entrepreneurs which plays an important role in the development of regional economies throughout Europe. The network ensures an A to Z range of assistance to new and existing SMEs: promotion, detection, selection, strategic support and post-launch follow-up.

Chapter 11
Perspectives

Abstract Bioinformatics has become a key part of modern life sciences research, and systems biology approaches are becoming more established. At the same time, the extreme complexity of living systems means that it is important to have longer-term perspectives to see how knowledge may be accumulated, interpreted and exploited as strategically and as usefully as possible, and to see what resources need to be developed to support these advances. A number of workshops have looked at perspectives for research in bioinformatics, systems biology and disease, and their insights are summarised here. The Seventh Framework Programme provides full flexibility for dealing with these areas, both in the short and in the longer term. Longer-term perspectives are also considered.

Introduction

Role of Perspectives

Bioinformatics has become a key part of modern life sciences research, and systems biology approaches are becoming more established. At the same time, the extreme complexity of living systems means that it is important to have longer-term perspectives to see how knowledge may be accumulated, interpreted and exploited as strategically and as usefully as possible, and to see what resources need to be developed to support these advances. This chapter concentrates on perspectives developed in a series of European Commission sponsored workshops listed in Table 1.1. Many of the perspectives discussed already have initial responses and actions in the most recent new projects started at the end of the Sixth Framework Programme (FP6) and the beginning of the Seventh Framework Programme (FP7), as discussed elsewhere in this book. These perspectives are therefore strongly based on the current state of the art and provide a framework for consideration of the future.

F.B. Marcus, *Bioinformatics and Systems Biology*,

DOI 10.1007/978-3-540-78353-4,

Bioinformatics

Bioinformatics Perspectives

The most relevant of the workshop reports on the strategic planning of bioinformatics requirements is the EU Projects Workshop Report on Systems Biology (Jehensen and Marcus 2005). This report also updates recommendations from the bioinformatics workshop report (Gyorffi and Marcus 2003). The content is also considered in terms of very recent books and studies (Nagl 2006 Lengauer 2007). In order to advance systems biology approaches, and also to develop more standard biological approaches to research, it is essential to develop underpinning bioinformatics tools and databases, and experimental support facilities, including tools and services: bioinformatics tools, databases, software, access, services, research and infrastructures as the basis for general biology research and for systems biology: It is recognised and emphasised that the full range of bioinformatics tools is a vital underpinning of all systems biology work. Bioinformatics analysis is essential in its own right and also in many cases merges smoothly with systems biology, depending on the nature of the problem. Resources for the future should include the following.

Model Systems and Biobank Resources

Research projects should reference well-characterised model systems with corresponding biobank resources, at the single-cell level, while linking these to multicellular model organisms and human and relevant biobank and tissue resources to develop aspects of health research. Potential single-cell model systems to be analysed include *Saccharomyces cerevisiae and Schizosaccharomyces pombe* (yeasts), *Bacillus subtilis, Escherichia coli* and *Lactococcus lactis* (filamentous fungi). Multicellular model organisms could include any of the standard model organisms, depending on data available, plus the human cells and cell lines relevant to particular health aspects; for example, mouse, rat, zebrafish, worm, *Arabidopsis*, mosquito, fly and various human cells, including neurons, hepatocytes and heart. Organisms are sometimes chosen for providing special contributions to either fundamental knowledge or specific health or biotechnological applications, but are often highly relevant to both. The analysis of human materials (e.g. tumour samples from cancer patients) helps provide a focus for analysis of specific disease processes.

Standards and Ontologies

Standards needed include those for data collection from experiments, storage in databases, "modelbases" and analysis, consistent with modelling requirements, including dynamic data where possible. This process is already under way with a wide range of data at the bioinformatics level, and needs to be extended to make the data useful for systems biology analysis, making full use of standard ontologies and controlled vocabularies and developing new ones where appropriate. Standards need to evolve with advances in systems biology. In collaborative research especially, standard operating procedures among partners are essential.

Obtaining Data Within and Beyond Present "Omics", Extending Databases

Although experimental biological data should be collected with bioinformatics and systems biology analysis in mind, there are key types of quantitative data becoming available which especially support systems biology, and which require special attention for standardisation and analysis. Current bioinformatics databases also need to be extended in order to make them compatible with the new systems biology information needs and expanded to take account of new information in these areas, including key areas for bioinformatics-based research and databases. In addition to extensive high-throughput data, there is also a need for accurate quantitative measurements in key areas to provide constraints (or very few constraints) to actual dynamic models.

The key data areas include:

- Gene expression, transcription, post-transcriptional control, regulatory RNAs
- Protein–protein, protein–nucleic acid and protein–metabolite interactions
- Protein modification
- Kinetics and non-equilibrium thermodynamics
- Genetic analysis and mutations
- Comparative genomics
- Metabolic flux analysis, metabolomics, lipidomics
- Alternative transcript and splicing data
- Control analysis
- In vivo imaging
- Haplotypes
- Protein homology
- Protein identification
- Protein structure
- Transcriptomics
- Toponomics
- Immunology
- Molecular concentrations/states (e.g., phosphorylation read-outs, etc.)

Systems Biology

Systems Biology Perspectives

The most relevant of the workshop reports on the strategic planning of topical areas is that by Jehensen and Marcus (2005), which also updates Marcus et al. (2004), plus recent reports as discussed later in the chapter (SYSBIOMED 2007). There are clear needs for mathematical modelling, concepts, principles and developing new methods for system identification, parameter estimation and spatio-temporal modelling. This also includes the combination of dynamic pathway models, formal analysis of the role of feedback in biochemical networks, their robustness and sensitivity as well as hybrid approaches to model the coordination cell function. Standardisation is necessary at the level of the networks and components modelled, model description (including reaction specification, measurement units, etc.), data storage and retrieval, and the computer codes. The last takes a particular importance because models developed by different networks/groups represent modules of cellular operation that should be compatible with each other. Hence, a priority is the development and use of multiplatform, non-proprietary programming languages, such as SBML (2007), with professional standards for software production and maintenance, and for Web-accessible live models (such as in the silicon cell). The ultimate goals are modular combinations of models and routine applications of "standard" models in non-specialist (experimental) laboratories.

Developing Systems Biology: Cellular and Subcellular Systems

A number of dedicated projects could be established to build up systems biology capabilities at the cellular and subcellular level. Research projects should focus on (1) modelling at least one process comprehensively, which is already a challenge, and where appropriate, by using levels of complexity to treat interactions, (2) integrated modelling of several cellular processes leading to as complete an understanding as possible of the dynamic behaviour of a cell or a tissue. Several projects may be required to develop modules (metabolism, signalling, trafficking, organelles, cell cycle, gene expression, replication, cytoskeleton) in model organisms. This modelling should involve realistic analysis of experimental data, including a wide range of data for transcriptomics, proteomics and functional genomics, and interactions with cellular pathways including signal transduction, regulatory cascades, metabolic pathways, etc. It should further involve generation and analysis of:

- Coherent, high-quality, quantitative, heterogeneous and dynamic experimental datasets as a basis for novel model constructions to advance from descriptive to predictive modelling
- The results from experimental functional analysis tools (in situ proteomics, protein–protein interactions, metabolic fluxes, etc.)

- Normal and diseased (perturbed) states, physiological and pathophysiological processes and their mechanisms and progression

The new FP7 (2007) systems biology projects discussed at the end of Chap. 3 in fact already are addressing several of these areas.

Multiple Interacting Systems at the Cellular and Physiological Levels

Much systems biology work is currently limited to specific levels of modelling and data integration. There is a need for improved technologies and methods for large-scale modelling, and current projects provide the basis for this advance.

- The applications will be wide-reaching, e.g. to develop the cell-level modelling to couple to the type of physiological modelling occurring in BIOSIM (2007), which directly aims at the drug design process. Understanding the hierarchical relationships will clearly allow a greatly improved description of function.
- Several international efforts have been launched, the metabolome, regulome, transcriptome, etc., to address functions at higher levels by mapping the nodes and networks involved in cellular and biological functions. These efforts will allow European laboratories to structure their efforts and contribute significantly to these international initiatives.
- This research demands a multidisciplinary approach linked to experimental work and data collection from high-throughput and emerging high-precision technologies, including array technologies, proteomics, molecular biology, bio-imaging and genetics of model organisms. A strong experimental component closely linked to the bioinformatics and computational systems biology component will be required for analysing and integrating the data collected. This multi-disciplinarity can often best be accommodated at the European level. Perhaps the most important next step is to properly integrate pathway-level systems biology research into a more complete model of interacting cellular functions.

The Europhysiome (2007) effort is based on a number of worldwide, collaborative Physiome (2007) project initiatives. The Physiome project is a worldwide public domain effort to provide a computational framework for understanding human and other eukaryotic physiological function. It aims to develop integrative models at all levels of biological organisation, from genes to the whole organism via gene regulatory networks, protein pathways, integrative cell function, and tissue and whole organ structure–function relations. Current projects include the development of:

- Ontologies to organise biological knowledge and access to databases
- Markup languages to encode models of biological structure and function in a standard format for sharing between different application programs and for reuse as components of more comprehensive models
- Databases of structure at the cell, tissue and organ levels

- Software to render computational models of cell function such as ion channel electrophysiological function, cell signalling and metabolic pathways, transport, motility, the cell cycle, etc. in two- and three-dimensional graphical form
- Software for displaying and interacting with the organ models which will allow the user to move across all spatial scales

Physiology

A key goal of health research is to model both the healthy state of human systems and the role of a disease mechanism and defence mechanisms, including the immune system. All these programmes will require resources and produce outputs that could be common to all of them, including databases, software tools, cellular and physiological models. The goal is understanding the functioning of physiological and pathological systems, involving the following:

- Sequence, structure, function, etc. bioinformatics databases, tools, research to support systems biology approaches
- Comprehensive model of a single cell
- Comprehensive model of physiological systems linking to human cells
- Neuroinformatics
- Modelling of the evolving state of organisms and people, involving developmental biology, ageing, mutations, circadian rhythms, etc.

The research being carried out by the projects described in Chaps. 7 and 8 illustrates the current state of the art in attempts to deal with these questions, but it is still just a beginning.

Biotechnology

Systems biology will be developed and applied in different areas such as the engineering or breeding of industrially important microorganisms and plants. Today, 25% of all medicines are plant-derived, and the spectrum of medicinal feedstocks and the efficiency of production can still be greatly enhanced. Projects such as AGRON-OMICS (2007), BaSysBio (2007) and BACELL-HEALTH (2007) illustrate progress already made towards these goals. Expertise developed in these areas may be beneficial for systems biology in medicine and vice versa and relevant coordination will be important. Systems biology also holds great potential to foster sustainable development, by accelerating and rationalising the production of plant-derived biofuels and chemical feedstocks in preparation of the inevitable depletion of fossil carbon.

Synthetic Biology

New applications of systems biology are being developed, as described in the Synthetic-Biology (2007) report, commissioned in the FP6 (2007) NEST new and emerging science and technology programme. Systems biology operates here at the interface with nanotechnology (Nano2 Life 2007). In FP6, NEST was used as an instrument to develop such emerging technologies.

Disease

Applying Systems Biology to Disease, Medicines and Treatment

Closely following the understanding of physiological processes is the need to model pathophysiological processes and diseases. In particular, systems biology will support the development and application of the medicines and various forms of treatment. The fields of pharmacokinetics and pharmacodynamics have long been established, but need to be taken to a much higher level of sophistication and simulation ability, as is being attempted in the BIOSIM (2007) project. In the cases of complex diseases such as the various types of cancer, it is essential to develop new strategies, based on the application of high-throughput techniques from functional genomics, to acquire information on most or all genes and gene products involved in the disease process, as well as in the response of the entire organism to any possible treatment. Analysing the very large quantity of information combining clinical, experimental and computational inputs requires the use of whole genome modelling to be able to translate information into predictions of the effects of different therapeutic schedules and drugs that are adapted to different specific genetic and physiological backgrounds. Systems biology could also potentially be applied to develop novel treatments and disease prevention regimes that reduce or circumvent the use of drugs. Areas include:

- Rationalised drug development and modelling; network-based drug design
- Improved modelling of effects, optimum dose and optimum timing of use of medicines, including personalised patient and group data where possible and appropriate
- Modelling of side effects of medicines, by modelling a wide range of systems in addition to the system targeted by the drug
- Modelling of altered nutrition and other life-style changes to treat or prevent pathological phenotypes
- Systems biology of complex, multifactorial diseases, e.g. varieties of cancer
- Modelling of the immune system, evolution of viral and bacterial resistance to medicines, interaction between organisms (internal ecology)
- Dynamical diseases (physiological disorders involving an abrupt switch to altered modes of dynamic behaviour)
- Public health relevant systems

Systems Biology of Targeted Diseases

SYSBIOMED (2007) seeks to explore the potential application of systems biology to medical research, including the development of drugs and other therapies. It is doing this through a series of workshops focusing on topics at the cutting edge of systems biology and physiology, which are being organised by young scientists working in relevant areas. Medical systems biology needs to demonstrate the ability to cross levels: from cells to organs and organisms, from cell function to physiological phenomena, and from model organisms to human diseases. Pioneering studies in the modelling of whole organ function have already demonstrated that models can correctly predict certain physiological and pathological functions of the heart. The first SYSBIOMED (2007) workshop agreed on the following priority areas suitable for a systems biology approach:

- Diabetes, which requires a systems approach
- Basal ganglia disorders, which cover a range of relevant pathological disorders, including Parkinson's disease, Huntington's chorea, schizophrenia, attention-deficit hyperactivity disorder, Tourette syndrome, drug addiction
- Regulation of inflammatory gene expression, since inflammation is associated with problems that relate to various diseases and allows the study of environmental effects (e.g. epigenetics of T and B cells)
- Apoptosis and cell cycle – chronotherapy, i.e. very important for understanding and optimising chemotherapies
- From networks to cell Fate – colon cancer: a problem for which existing knowledge, data and experimental systems provide an ideal basis to explore systems biology in a translational effort
- Single-cell technologies: indispensable tool for generation of high-value data on cellular physiological processes, includes phosphoproteomics; from images to models
- Model integration: the larger picture requires methods to identify (to define the boundaries of) subsystems (modules), explore these in experiments and then integrate models thereof

Therapeutic Applications of Computational Biology and Chemistry

SYSBIOMED (2007) complements both ongoing and planned European systems biology initiatives. The EBI (2007) organised a workshop on the therapeutic applications of computational biology (TACB 2005). A detailed background is provided by papers presented at TACB (2005). They note that computational biology is now of key strategic importance to the pharmaceutical industry; it is used at all stages of the drug discovery and development pipeline, from target identification through to regulatory approval. Computational approaches to the search for new therapeutic, diagnostic and preventive approaches to disease embrace not only bioinformatics, but also cheminformatics and medical informatics. The basis of longevity was discussed using information on genetic variations that are common only to those people who have lived beyond the age of 90. Informatics-based approaches were considered for predicting properties such as toxicity or poor bioavailability (the extent to which a drug is available to the tissue) that can cause otherwise promising lead compounds to fail. A combination of sequence- and structure-based knowledge of drug targets, combined with information about how proteins bind their cognate ligands, can be used to assess the likelihood that targets can be affected by drugs. This approach has led to the conclusion that only a small proportion (perhaps as little as 1%) of the human genome can be targeted by drugs. The use of microarray-based data to predict toxicity was the focus of the toxicoinformatics session. Statistical learning methods can be used to predict the emergence of drug resistance in patients with HIV/AIDS, and thereby to optimise their treatment regimen. The virtual cancer patient model provides a computational approach to predicting disease progression and optimising drug combinations and schedules by simulating the dynamics of key processes underlying drug–patient interactions. It allows drug developers to perform numerous rapid virtual clinical trials and to forecast optimal drug treatments. The second TACB conference (TACB 2007) expanded on these themes and the challenges therein.

Seventh Framework Programme

Health Research

The European Commission FP7 (2007) programme will fund new projects starting between 2007 and 2013, each lasting typically 3–5 years. Bioinformatics and systems biology and supporting technologies will receive significant support within this programme. Funding supports research and also links together existing efforts in Europe, by building on the strong base already established. Details are available in Chap. 10.

Physiological Integration

In a highly ambitious vision of systems biology, the Europhysiome (2007) consortium proposes the Virtual Physiological Human, which is a methodological and technological framework that if established would enable the investigation of the human body as a single complex system. The consortium argues that the human body is like a jigsaw puzzle made of a trillion pieces. Currently people try to understand the whole picture by looking at a single piece, or at a few closely interconnected pieces. The Virtual Physiological Human is the frame within which the consortium proposes to place the pieces all together, and the glue that connects them. Implementation of this vision has been started by the European Commission DG Information and Communication Technologies, with a call for proposals FP7-ICT-2007-2 (2007) including "Objective ICT-2007.5.3: Virtual Physiological Human."

Longer Term

Systems Biology Forward Look

The European Science Foundation (ESF 2007) published a forward look paper on systems biology (ESF-SysBio 2007), and has organised a task force which has produced a detailed position paper (ESF-Task-Force 2007). This ESF Task Force on Systems Biology sets out a road map and specific recommendations intended to establish a world leading systems biology research programme in Europe. The recommendations are grounded in the ESF "Forward Look on Systems Biology", and on a broad strategic overview provided by the task force.

Very Long Term

When looking at long-term perspectives, the choice of breadth and time is key. For a broad sweeping vision of a possible long-term future, Kaku (1998) attempted to project the field of biology to 2050 and beyond. Concerning those visions, perhaps we should not do everything we are capable of doing technologically! Such perspectives raise major ethical questions. Even in the present, there is a European Commission ethical screening programme is in place (FP7-ethics 2007) for proposals. Still, great progress in understanding and in developing cures for diseases will undoubtably be made.

Chapter 12
Outstanding Results and Conclusions

Abstract A number of selected outstanding results and new resources due to the European Commission funded collaborative projects are summarised. These projects are often crucial to understanding and advances in a particular field. These results emphasise that collaborative research has had major successes, and that in many ways it has transformed life science research. Impact is judged in terms of greatly increasing the capabilities of biomedical researchers around the world to carry out their research, and in terms of major contributions to fields of research, often with direct application to the understanding and treatment of diseases. Overall conclusions of the book are presented.

Introduction

Tests of the Value of Collaborative Research

This book has claimed that the collaborative research paradigm has created new possibilities and produced results in a way that ordinary research either could not or would be much less efficient in doing so. In contrast, it has been argued that collaborative research is equivalent to funding many individual researchers at a range of laboratories with common interests, and that similar results could be obtained without complicated contracts and management structures by just giving the money to the laboratories. Additionally, many informal collaborations, especially those based on an open source model, are able to create impressive resources by pooling data and tools. Therefore, one test is to look at the results of such projects, and to compare with either competitors or the situation before the project existed, and to ask if the collective results of the projects have made major contributions to the scientific community in terms of contributions and resources. On the basis of the contents of this book and the examples shown in this chapter, the answer would seem to be yes.

F.B. Marcus, *Bioinformatics and Systems Biology*,
DOI 10.1007/978-3-540-78353-4,

The Impact of Collaborative Research

To emphasise that collaborative research has had major successes, and that in many ways it has transformed life science research, a number of selected outstanding results and new resources due to the European Commission funded collaborative projects are summarised here, which are often world-leading or crucial to understanding and advances in a particular field. Here, the impact is judged in terms of greatly increasing the capabilities of biomedical researchers around the world to carry out their research, and in terms of major contributions to fields of research, often with direct application to the understanding and treatment of diseases.

Outstanding Resources Created

Integrated Data Access Capabilities

In the past, the first website usually accessed by life science researchers was Entrez (2007), leading to PubMed (2007) at the NCBI (2007) in the USA. Currently via its homepage, the EBI (2007) has recently vastly improved its capabilities, by including EB-eye and highly linked and cross-referenced databases and tools, as a result of work done in TEMBLOR (2007), BioSapiens (2007) and EMBRACE (2007). These capabilities are already leading to rapidly increasing numbers of worldwide users.

Establishment of Standards and Repository for Gene Expression Data

The project DESPRAD (2007) and subsequent support from BioSapiens (2007) and the collaborative support of many European teams has created world-recognised standards and data repositories in ArrayExpress (2007), which is the reference for the central technology of DNA chip measurements of gene expression data.

A European Bioinformatics Grid and Linked Resources

Although there is already access via the Internet to a very wide range of databases, the bioinformatics grid being established by EMBRACE (2007) is transforming bioinformatics and systems biology. Although data around the world are accessible via the Internet, this access is severely hampered by different database formats and protocols, and the difficulty of extensive data processing that involves more than one database. Grid capabilities for data and computing had already been established via European and national grid hardware and middleware networks (e.g. the GEANT2 2007 and EGEE 2007 grid projects funded by the European Commission Directorate General for the Information Society). These capabilities include links to worldwide grid networks. However, detailed Web software, Web services and protocols for information presented externally by bioinformatics databases needed to be established to make use of this capability. This is what is being accomplished by the EMBRACE (2007) grid. Its scientific test cases show that this technology approach is transforming the way biomedical research is approached in a wide variety of key problems, for both distributed data access and distributed computing.

European Virtual Institute of Genome Annotation

In the WTEK survey (Cassman et al. 2007), it was concluded that the European Virtual Institute of Genome Annotation established by BioSapiens (2007) had capabilities that were unequalled elsewhere in the world. Genome annotation in this context is much more than assigning a single ontology term to a single gene. As shown in the BioSapiens (2007) reports, annotation involves characterising most of the database-stored knowledge in biology. Knowledge is annotated and linked from the beginning at gene identification through to protein sequence and structure to protein–protein interactions and networks to integrated applications to directly applicable disease research questions. The DAS (2007) distributed annotation systems have been extended to most domains of bioinformatics to provide worldwide capabilities of both contributing data and viewing all data in an integrated way via Ensembl (2007).

European School of Bioinformatics and Training and Outreach

There are a very wide range of university degree and training courses, workshops and conferences, every year, sponsored by a wide range of institutions. However, the European School of Bioinformatics established by BioSapiens (2007) provides unique capabilities in that it takes students from all over Europe, and it directly connects students with the latest results from the European Virtual Institute of Genome annotation, by providing basic training for users of all bioinformatics tools, and more advanced training on each of the topics of the research agenda, symposia on biological systems of interest for the network and ensuring that postdoctoral workers hired by the network are trained effectively. The School is also fostering a collaborative effort among master's degree programmes to try and provide useful information to potential students and to help coordination among the various initiatives.

Outstanding Scientific Results

The Majority of Human DNA Is Transcribed into RNA

A major paper describing the work of the NIH (2007) funded ENCODE (2007) consortium and a number of accompanying papers have been published by the ENCODE-Project-Consortium (2007). The BioSapiens (2007) results are available at GENCODE (2007), see Tress et al. (2007). The findings of ENCODE-Project-Consortium (2007) promise to reshape our understanding of how the human genome functions. They challenge the traditional view of our genetic blueprint as a tidy collection of independent genes, pointing instead to a network in which genes, regulatory elements and other types of DNA sequences interact in complex, overlapping ways. In an analysis effort led by the European partners from BioSapiens (2007), the ENCODE (2007) consortium's major findings include the discovery that the majority of human DNA is transcribed into RNA and that these transcripts extensively overlap one another. This broad pattern of transcription challenges the long-standing view that the human genome consists of a small set of discrete genes, along with a vast amount of "junk" DNA that is not biologically active. The new data indicate that the genome contains few unused sequences; genes are just one of many types of DNA sequences that have a functional impact. These discoveries are fundamental to the future course of biomedical research, even recognised as a Newsweek cover story (Silver 2007).

Understanding the Dynamic Behaviour of the p53–Mdm2 Network

The p53 gene has perhaps the most publications devoted to it, currently around 44,000. In the COMBIO (2007) project, the experimental work of Geva-Zatorsky et al. (2006) employed the negative-feedback loop between the tumour-suppressor p53 and the oncogene Mdm2. They found that isogenic cells in the same environment behaved in highly variable ways following DNA-damaging γ-irradiation: some cells showed undamped oscillations for at least 3 days (more than ten peaks). The amplitude of the oscillations was much more variable than the period. They also analysed different families of mathematical models of the system, including a novel checkpoint mechanism. The models pointed to the possible source of the variability in the oscillations: low-frequency noise in protein production rates, rather than noise in other parameters such as degradation rates. A mathematical model (Ciliberto et al. 2005) of these p53 oscillations based on positive and negative feedbacks in the p53–Mdm2 network indicated that the system reacts to DNA damage by moving from a stable steady state into a region of stable limit cycles. Oscillations in the model are born with large amplitude, which guarantees an all-or-none response to damage. As the p53 concentration oscillates, damage is repaired and the system moves back to a stable steady state with low p53 activity. The model reproduces experimental data in quantitative detail.

Understanding the Dynamics of Spindle Formation in Cells

Measurements of gradients, time-lapse image acquisition of mitotic events, green fluorescent protein localisation, etc. have been done by the COMBIO (2007) experimental groups and resulted in the generation of a life imaging repository (Rebollo et al. 2007) of spindle assembly movies in different species and in different cell lineages within a species. Using these data and those compiled from the literature, the theoretical groups have developed models and performed simulations to account for the role of chromatin, phosphorylation gradients and component localisation in microtubule (MT) nucleation and organisation. Predictions made from the simulations were tested experimentally (Janson et al. 2007). They concluded that MT nucleation not only occurs from centrosomes, but also in large part from dispersed nucleation sites. The subsequent sorting of short MTs into networks like the mitotic spindle requires molecular motors that laterally slide overlapping MTs and bundling proteins that statically connect MTs. How bundling proteins interfere with MT sliding was unclear. In bipolar MT bundles in fission yeast, they found that the bundler ase1p localised all along the length of antiparallel MTs, whereas the motor klp2p (kinesin-14) accumulated only at MT plus ends. Consequently, sliding forces could only overcome resistant bundling forces for short, newly nucleated MTs, which were transported to their correct position within bundles. Ase1p thus regulated sliding forces on the basis of polarity and overlap length, and computer simulations showed these mechanisms to be sufficient to generate stable bipolar bundles. By combining motor and bundling proteins, cells can thus dynamically organise stable regions of overlap between cytoskeletal filaments.

Establishment of the Role of Alternative Transcription and Splicing in Cancer

Identification of cancer-specific splice forms is a major objective in genomic medicine because these transcripts are efficient cancer signatures and could constitute drug targets. In ATD (2007), verification of alternative transcript cancer markers was performed by reverse transcription (RT) PCR techniques, using cell lines derived from neoplastic human tissues of the colon, the cervix and the lung. ATD (2007) uses a novel bioinformatics approach that selects candidate disease genes according to their alternative transcript expression profiles. It uses the anatomical eVOContology (2007) to mine available human gene expression data for cancer-specific events.

To demonstrate that the method is successful and widely applicable, 424 splice events were chosen for RT-PCR experiments: 230 candidate splice events predicted to be cancer-related and 194 corresponding reference events which should be detectable in normal and/or neoplastic cells. The experiments showed a rather cancer-related expression for 32% of candidate splice events comparing normal human tissues and human cancer cell lines (Gautheret 2007). This approach facilitates direct association between genomic data describing gene expression and information from biomedical texts describing disease phenotype, and successfully prioritises candidate genes according to their expression in disease-affected tissues.

Small-Ligand Binding

Some of the major problems related to metabolism and drug development are related to the binding of small ligands to proteins. Stockwell and Thornton (2006) observed that the phenomenon of molecular recognition, which underpins almost all biological processes, is dynamic, complex and subtle. They presented an analysis of the conformational variability exhibited by three of the most ubiquitous biological ligands in nature, ATP, NAD and FAD, and demonstrated qualitatively that these ligands bind to proteins in widely varying conformations, including several cases in which parts of the molecule assume energetically unfavourable orientations. Several other results are presented that are fundamental to structure-to-function interpretations concerning proteins and ligands BioSapiens (2007).

Multiple Drug Treatment for HIV/AIDS Drug-Resistant Mutation Pathways

Important summaries of techniques used to predict optimum AIDS/HIV therapies are found at MPI-INF-Bioinformatics-for-HIV (2007), further developed as part of the BioSapiens (2007) project. Altmann et al. (2007) showed that the outcome of antiretroviral combination therapy depends on many factors involving host, virus and drugs. Predictions were made of treatment response from the applied drug combination and the genetic constellation of the virus population at baseline. The virus's evolutionary potential for escaping from drug pressure was explored as an additional predictor. Different encodings of the viral genotype and antiretroviral regimen were compared, including phenotypic and evolutionary information, namely predicted phenotypic drug resistance, activity of the regimen estimated from sequence space search, the genetic barrier to drug resistance and the genetic progression score. The benefit of phenotypic information in predicting virological response was confirmed by using predicted fold changes in drug susceptibility. Moreover, genetic barrier and predicted phenotypic drug resistance were found to be the best encodings across all datasets and statistical learning methods examined.

A prototypical implementation of the best performing approach is freely available for research purposes at Geno2pheno (2007). These results were discussed at the BioSapiens-viRgil-Workshop (2007).

Cancer as A Signalling Disease, Applied to Protein Kinase Pathways

In a discussion of mitogen-activated protein kinase (MAPK) signalling pathways in cancer (Dhillon et al. 2007), the COSBICS (2007) project shows areas in which cancer can be perceived as a disease of communication between and within cells. The aberrations are pleiotropic, but MAPK pathways feature prominently. Dhillon et al. (2007) discuss recent findings and hypotheses on the role of MAPK pathways in cancer. Cancerous mutations in MAPK pathways frequently mostly affect Ras and B-Raf in the extracellular signal-regulated kinase pathway. Stress-activated pathways, such as Jun N-terminal kinase and p38, largely seem to counteract malignant transformation. The balance and integration between these signals may widely vary in different tumours, but are important for the outcome and the sensitivity to drug therapy.

Future Project Outcomes

All of the Sixth Framework Programme projects discussed here started on or after January 2004, and some have only recently begun; therefore, the level of achievement in this relatively short time is already extremely high. Many of the projects are just entering their most productive phase, and even more impressive research is in progress (see Websites). The nature of the resources and results discussed above indicates that collaborative research, with close and continuous interaction between the participants, was in fact highly necessary for such results to have been achieved.

Overall Conclusions

The Challenge

In the conclusion to his book on the biology of cancer, Weinberg (2007) writes: "… Successes in these efforts, involving the new discipline of "systems biology," will surely benefit cancer research. Imagine a day – still years away – when the biological responses of various human cells, normal and malignant, can be predicted by mathematical models of these cells and their internal control circuits…. For now, at least, we need to wrestle with the grim realities of drug development, the inadequate animal models, our ignorance of the behaviour of cellular regulatory circuitry, and the confounding biological complexities of human cancer."

Successes of Collaborative Research in Bioinformatics and Systems Biology

The discussion in this book would seem to have demonstrated that important advances have already been made on the way to responding to these cancer-specific and other biology- and health-related challenges and to realising these aspirations. The steps taken are already valuable in terms of understanding and applicable results, especially in the context of the following overall conclusions:

- Major research communities in bioinformatics and systems biology have been created by collaborative research projects of the European Commission.
- These projects are strongly interlinked internally and externally, and their results, publications, new resources, tools and services provide major contributions to world research capabilities and results.
- These results and resources are available worldwide, by means of formal and informal collaboration, by Internet- and grid-based access, in the literature, and from the experience of the researchers.
- These collaborative scientific research paradigms are successful, and often result in unique and world-leading capabilities, and represent an excellent way in the future to approach a wide range of problems in the biomedical sciences.

References

ACGT (2007) Advancing clinicogenomic trials on cancer. http://eu-acgt.org/about-us/more-information/wp-description.html. Accessed 1 Dec 2007

AfCS (2007) Alliance for cell signalling. http://www.afcs.org. Accessed 1 Dec 2007

AGRON-OMICS (2007) Arabidopsis growth network integrating omics technologies. http://www.agron-omics.eu. Accessed 1 Dec 2007

AGRON-OMICS-Management (2007) AGRON-OMICS Consortium Structure http://www.agron-omics.eu/index.php/consortium/consortium_structure. Accessed 1 Dec 2007

Alberghina L, Westerhoff HV (eds) (2005) Systems biology – definitions and perspectives. Springer, Berlin

Alberts B, Johnson A, Lewis J, Raff M, Roberts K, Walt P (2002) Molecular biology of the cell, 4th edn. Garland, New York

Alon U (2006) An introduction to systems biology: design principles of biological circuits. Taylor & Francis, Boca Raton

Altmann A, Beerenwinkel N, Sing T, Savenkov I, Däumer M, Kaiser R, Rhee S-Y, Fessel WJ, Shafer RW, Lengauer T (2007) Improved prediction of response to antiretroviral combination therapy using the genetic barrier to drug resistance. Antivir Ther 12:169–178

Amico M, Finelli M, Rossi I, Zauli A, Elofsson A, Viklund H, Heijne G, von Jones D, Krogh A, Fariselli P, Martelli PL, Casadio R (2006) PONGO: a web server for multiple predictions of all-alpha transmembrane proteins. Nucleic Acids Res 34:169–172

AMPKIN (2007) Systems biology of the AMP-activated protein kinase. http://www.gmm.gu.se/AMPKIN. Accessed 1 Dec 2007

ANGIOTARGETING (2007) Tumour angiogenesis research. http://www.uib.no/med/angiotargeting/. Accessed 1 Dec 2007

ArrayExpress (2007) Public repository for microarray data http://www.ebi.ac.uk/arrayexpress/. Accessed 1 Dec 2007

ASD (2007) Alternative Splicing Database project. http://www.ebi.ac.uk/asd/. Accessed 1 Dec 2007

ASTD (2007) Alternative Splicing and Transcript Diversity database. http://www.ebi.ac.uk/astd. Accessed 1 Dec 2007

ATD (2007) Alternate Transcript Diversity. http://www.atdproject.org. Accessed 1 Dec 2007

Atrih A, Richardson J, Prescott AR, Ferguson MA (2005) Trypanosoma brucei glycoproteins contain novel giant poly-N-acetyllactosamine carbohydrate chains. J Biol Chem 280(2):865–871

ATTACK (2007) Adoptive engineered T cell targeting to activate cancer killing. http://www.attack-cancer.org. Accessed 1 Dec 2007

Attwood TK, Parry-Smith DJ (1999) Introduction to bioinformatics. Prentice-Hall, New York

BACELL-HEALTH (2007) Bacillus cell factory for EU bio industries. http://www.bacell.eu. Accessed 1 Dec 2007

Barcelona (2007) Department of clinical sciences – University of Barcelona. http://www.ub.edu/organitzacio/en/departments/endepclinicsciences.htm. Accessed 1 Dec 2007

BaSysBio (2007) Towards an understanding of dynamic transcriptional regulation at global scale in bacteria: a systems biology approach. http://www.basysbio.eu/. Accessed 1 Dec 2007

Baxevanis AD, Ouellette BFF (2001) Bioinformatics: a practical guide to the analysis of genes and proteins. Wiley-Interscience, New York

Beausoleil SA et al.(2004) Large-Scale characterization of HeLa cell nuclear phosphoproteins. Proc Natl Acad Sci USA 101:12130–12135

Beerenwinkel N, Schmidt B, Walter H, Kaiser R, Lengauer T, Hoffmann D, Korn K, Selbig J (2001) Geno2pheno: a new machine learning approach to predicting phenotypic drug resistance from genotype. Antivir Ther 6(Suppl 1):113

Bertau M, Mosekilde E, Westerhoff HV (2007) Biosimulation in drug development. Wiley, New York

BioBabel (2007) Enhanced interoperability of biological databases by standardisation of biochemical terminology and introduction of a shared ontology. http://www.ebi.ac.uk/biobabel/. Accessed 1 Dec 2007

Biobanks (2005) From biobanks to biomarkers – translating the potential of human population genetic research to improve the quality of health of the European Union citizen. Proceedings of a conference held at the Wellcome Trust Conference Centre, Hinxton, Cambridge, 20–22 September 2005. http://www.wellcome.ac.uk/stellent/groups/corporatesite/sitestudioobjects/documents/web_document/wtx032086.pdf. Accessed 1 Dec 2007

BioBase (2007) BioBase biological databases. http://www.biobase.de. Accessed 1 Dec 2007

BioBridge (2007) Integrative genomics and chronic disease phenotypes: modelling and simulation tools for clinicians. http://www.biobridge.eu. Accessed 1 Dec 2007

BioCyc (2007) Pathway/genome databases. http://www.biocyc.org/. Accessed 1 Dec 2007

BioMap (2007) Functional and structural resources for bioinformatics. http://www.biochem.ucl.ac.uk/bsm/biomap/. Accessed 1 Dec 2007

BioMart (2007) Query-oriented data management system. http://www.biomart.org. Accessed 1 Dec 2007

BioMinT (2007) Biological text mining. http://www.phamadm.com/biomint/. Accessed 1 Dec 2007

BioModels (2007) A database of annotated published models. http://www.ebi.ac.uk/biomodels/. Accessed 1 Dec 2007

BioSapiens (2005) BioSapiens: a European network for integrated genome annotation. Eur J Hum Genet 13:994–997

BioSapiens (2007) A European virtual institute for genome annotation. http://www.biosapiens.info. Accessed 1 Dec 2007

BioSapiens-Management (2007) BioSapiens management structure. http://www.biosapiens.info/page.php?page=management. Accessed 1 Dec 2007

BioSapiens-partners (2007) BioSapiens list of partners. http://www.biosapiens.info/page.php?page=partners&sid=6e1bb8bbbf93acb7266f16d9ce45a85c. Accessed 1 Dec 2007

BioSapiens-Publications (2007) BioSapiens list of project publications. http://www.biosapiens.info/page.php?page=publications. Accessed 1 Dec 2007

BioSapiens-viRgil-Workshop (2007) BioSapiens–viRgil workshop on bioinformatics for viral infections. http://workshop2005.bioinf.mpi-sb.mpg.de. Accessed 1 Dec 2007

BioSapiens-WP1 (2007) BioSapiens work package 1 – gene definition/alternative splicing. http://www.biosapiens.info/page.php?page=package&pack=1. Accessed 1 Dec 2007

BioSapiens-WP2 (2007) BioSapiens work package 2 – regulators and promoters. http://www.biosapiens.info/page.php?page=package&pack=2. Accessed 1 Dec 2007

BioSapiens-WP3 (2007) BioSapiens work package 3 – expression. http://www.biosapiens.info/page.php?page=package&pack=3. Accessed 1 Dec 2007

BioSapiens-WP4 (2007) BioSapiens work package 4 – variation – haplotypes and SNPs. http://www.biosapiens.info/page.php?page=package&pack=4. Accessed 1 Dec 2007

BioSapiens-WP5 (2007) BioSapiens work package 5 – protein families, othologues. http://www.biosapiens.info/page.php?page=package&pack=5. Accessed 1 Dec 2007

BioSapiens-WP6 (2007) BioSapiens work package 6 – membrane proteins and ligands. http://www.biosapiens.info/page.php?page=package&pack=6. Accessed 1 Dec 2007

BioSapiens-WP7 (2007) BioSapiens work package 7 – 3D protein structure. http://www.biosapiens.info/page.php?page=package&pack=7. Accessed 1 Dec 2007

BioSapiens-WP8 (2007) BioSapiens work package 8 – post-translation modification and localisation. http://www.biosapiens.info/page.php?page=package&pack=8. Accessed 1 Dec 2007

BioSapiens-WP9 (2007) BioSapiens work package 9 – sequence and structure to function. http://www.biosapiens.info/page.php?page=package&pack=9. Accessed 1 Dec 2007

BioSapiens-WP10 (2007) BioSapiens work package 10 – protein-protein complexes. http://www.biosapiens.info/page.php?page=package&pack=10. Accessed 1 Dec 2007

BioSapiens-WP11 (2007) BioSapiens work package 11 – pathways and networks. http://www.biosapiens.info/page.php?page=package&pack=11. Accessed 1 Dec 2007

BioSapiens-WP15 (2007) BioSapiens work package 15 – onfectious diseases. http://www.biosapiens.info/page.php?page=package&pack=15. Accessed 1 Dec 2007

BioSapiens-WP16 (2007) BioSapiens work package 16 – Down's syndrome. http://www.biosapiens.info/page.php?page=package&pack=16. Accessed 1 Dec 2007

BioSapiens-WP20 (2007) BioSapiens work package 20 – ENCODE. http://www.biosapiens.info/page.php?page=package&pack=20. Accessed 1 Dec 2007

BioSapiens-WP101 (2007) BioSapiens work oackage 101 – gene definition and alternative splicing. http://www.biosapiens.info/page.php?page=package&pack=101&new=1. Accessed 1 Dec 2007

BioSapiens-WP102 (2007) BioSapiens work package 102 – gene regulation and expression. http://www.biosapiens.info/page.php?page=package&pack=102&new=1. Accessed 1 Dec 2007

BioSapiens-WP103/110 (2007) BioSapiens work package WP 103/110 – variation (haplotypes and SNPs), incorporating the thematic work package on complex trait proteins. http://www.biosapiens.info/page.php?page=package&pack=103/110&new=1. Accessed 1 Dec 2007

BioSapiens-WP104 (2007) BioSapiens work package 104 – functional annotation of proteins. http://www.biosapiens.info/page.php?page=package&pack=104&new=1. Accessed 1 Dec 2007

BioSapiens-WP105 (2007) BioSapiens work package 105 – post-translation modification, membrane and localisation prediction. http://www.biosapiens.info/page.php?page=package&pack=105&new=1. Accessed 1 Dec 2007

BioSapiens-WP106 (2007) BioSapiens work package 106 – protein complexes, networks and pathways. http://www.biosapiens.info/page.php?page=package&pack=106&new=1. Accessed 1 Dec 2007

BioSapiens-WP108 (2007) BioSapiens work package 108 – ENCODE. http://www.biosapiens.info/page.php?page=package&pack=108&new=1. Accessed 1 Dec 2007

BioSapiens-WP109 (2007) BioSapiens work package 109 – cancer. http://www.biosapiens.info/page.php?page=package&pack=109&new=1. Accessed 1 Dec 2007

BIOSIM (2007) Biosimulation – a new tool in drug development http://biosim.fysik.dtu.dk:8080/biosim. Accessed 1 Dec 2007

BMIS (2007) British Medical Informatics Society. http://www.bmis.org/. Accessed 1 Dec 2007

Bock G, Cohen D, Goode JA (eds) (2000) From genome to therapy: integrating new technologies with drug development. Novartis Foundation Symposium 229, Wiley, Chichester, UK

Bock G, Goode JA (eds) (2002) *'In silico'* simulation of biological processes. Novartis Foundation Symposium 247, Wiley, Chichester, UK

BRECOSM (2007) Identification of molecular pathways that regulate the organ-specific metastasis of breast cancer. http://igtmv1.fzk.de/www/brecosm/. Accessed 1 Dec 2007

BRENDA (2007) The comprehensive enzyme information system. http://www.brenda-enzymes.info/. Accessed 1 Dec 2007

Bringmann P, Butcher E, Parry G, Weiss B (eds) (2007) Systems biology – applications and perspectives series: Ernst Schering Foundation symposium proceedings, 61Springer, Berlin

CABIG (2007) Cancer biomedical information grid (caBIG™) of the NCI (2007). https://cabig.nci.nih.gov/. Accessed 1 Dec 2007

Cancer Genome Project (2007) The WTSI (2007) cancer genome project. http://www.sanger.ac.uk/genetics/CGP/. Accessed 1 Dec 2007

Carlson BM (2004) Human embryology and developmental biology, 3rd edn. Elsevier/Mosby, Philadelphia

Carro A, Tress M, Juan D, de Pazos F, Lopez-Romero P, Sol A, del Valencia A, Rojas AM (2006) TreeDet: a web server to explore sequence space. Nucleic Acids Res 1:34

Cassidy J, Bissett D, Spence RAJ (2002) Oxford handbook of oncology. Oxford University Press, Oxford

Cassman M, Arkin A, Doyle F, Katagiri F, Lauffenburger D, Stokes C (2007) Systems biology – international research and development. Springer, Berlin

Cassman M, Brunak S (2007) The US-EC workshop on infrastructure needs for systems biology. http://bnmc.caltech.edu/doku.php?id=us-ec-workshop and http://ec.europa.eu/research/biotechnology/ec-us/index_en.html. Accessed 1 Dec 2007

CATH (2007) Protein structure classification database. http://www.cathdb.info/latest/index.html. Accessed 1 Dec 2007

CBS-DTU (2007) Center for Biological Sequence Analysis – Danish Technical University. http://www.cbs.dtu.dk/. Accessed 1 Dec 2007

CBS-DTU-Biolinks (2007) CBS-DTU (2007) links to bioinformatics capabilities. http://www.cbs.dtu.dk/biolinks/. Accessed 1 Dec 2007

CCO (2007) Cell cycle ontology. http://www.cellcycleontology.org. Accessed 1 Dec 2007

Celera (2007) An Applera Corporation business. http://www.celera.com. Accessed 1 Dec 2007

Cells-into-organs (2007) Cells into organs: functional genomics for development and disease of mesodermal organ systems. http://www.cellsintoorgans.net. Accessed 1 Dec 2007

CDL (2007) Cell division laboratory. http://www.pcb.ub.es/divisioncelular/Publications.html. Accessed 1 Dec 2007

CERM (2007) Magnetic resonance center. http://www.cerm.unifi.it. Accessed 1 Dec 2007

CIB-DDBJ (2007) Centre for Information Biology and DNA Data Bank of Japan. http://www.cib.nig.ac.jp/. Accessed 1 Dec 2007

Ciliberto A, Novak B, Tyson JJ (2005) Steady states and oscillations in the p53/Mdm2 network. Cell Cycle 4(3):488–493

CIS-modules (2007) Conditions for putative cis modules. http://sysdb.cs.helsinki.fi/u/tkt_bsap/EELweb/. Accessed 1 Dec 2007

CiteXplore (2007) Literature search with text mining tools. http://www.ebi.ac.uk/citexplore. Accessed 1 Dec 2007

Claverie J-M, Notredame C (2003) Bioinformatics for dummies. Wiley, New York

CluSTr (2007) Automatic classification of UniProt Knowledgebase and IPI proteins into groups of related proteins. http://www.ebi.ac.uk/clustr/. Accessed 1 Dec 2007

CMB-DTA (2007) Center for Microbial Biotechnology – Danish Technical University. http://www.cmb.dtu.dk/English.aspx. Accessed 1 Dec 2007

COMBIO (2007) An integrative approach to cellular signalling and control processes: Bringing computational biology to the bench. http://combio.crg.es. Accessed 1 Dec 2007

CORDIS (2007) European Community Research and Development Information Service. http://cordis.europa.eu/en/home.html. Accessed 1 Dec 2007

CORDIS-Projects (2007) CORDIS – find a project. http://cordis.europa.eu/fp6/projects.htm. Accessed 1 Dec 2007

COSBICS (2007) Computational systems biology of cell signalling http://www.sbi.uni-rostock.de/cosbics. Accessed 1 Dec 2007

COSMIC (2007) Catalogue of somatic mutations in cancer. http://www.sanger.ac.uk/genetics/CGP/cosmic. Accessed 1 Dec 2007

CPA (2007) Center for proteome analysis. http://www1.sdu.dk/health/research/units/proteomeana.php. Accessed 1 Dec 2007

CRESCENDO (2007) Consortium for research into nuclear receptors and ageing. http://www.crescendoip.org. Accessed 1 Dec 2007

CRESCENDO-links (2007) Links from consortium for research into nuclear receptors and ageing http://www.crescendoip.org/links.asp?rub=4. Accessed 1 Dec 2007

Crespi S (2001) Managing intellectual property rights in a knowledge-based economy – bioinformatics and the influence of public policy. ftp://ftp.cordis.europa.eu/pub/life/docs/ipr_bioinf.pdf. Accessed 1 Dec 2007

CRG (2007) Centre for Genomic Regulation – design of biological systems. http://www.crg.es/luis_serrano. Accessed 1 Dec 2007

Curwen V et al.(2004) The Ensembl analysis pipeline. Genome Res 14:942–950

CVIT (2007) The Center for the Development of a Virtual Tumour. https://www.cvit.org. Accessed 1 Dec 2007

Dale JW, Park SF (2004) Molecular genetics of bacteria. Wiley, Chichester

DAS (2007) Distributed annotation system. http://www.biosapiens.info/page.php?page=das_portal&sid=683482e348c96177c1e0213cc0dbf750. Accessed 1 Dec 2007

DAS-Information (2007) DAS server information service. http://www.biosapiens.info/page.php?page=biosapiensdir. Accessed 1 Dec 2007.

DASTY (2007) The UNIPROT DAS tool. http://www.ebi.ac.uk/dasty. Accessed 1 Dec 2007

dbSNP (2007) Single nucleotide polymorphism database. http://www.ncbi.nlm.nih.gov/SNP. Accessed 1 Dec 2007

DECHEMA (2007) The Society for Chemical Engineering and Biotechnology. http://www.dechema.de/en/The+DECHEMA.html. Accessed 1 Dec 2007

DeCode (2007) deCODE genetics biopharmaceutical company http://www.decode.com. Accessed 1 Dec 2007

3D-EM (2007) Three dimensional electron microscopy. http://www.3dem-noe.org/. Accessed 1 Dec 2007

DESPRAD (2007) Development and establishment of standards and prototype repository for DNA-array data. http://www.ebi.ac.uk/microarray/Projects/desprad/. Accessed 1 Dec 2007

Dhillon AS, Hagan S, Rath O, Kolch W (2007) MAPK kinase signalling pathways in cancer. Oncogene 26(22):3279–3290

Disease Ontology (2007) Disease Ontology. http://diseaseontology.sourceforge.net/. Accessed 1 Dec 2007

Di Ventura B, Lemerle C, Michalodimitrakis K, Serrano L (2006) From in vivo to in silico biology and back. Nature 443:527–533

DIAMONDS (2007) Dedicated integration and modelling of novel data and prior knowledge to enable systems biology. http://www.sbcellcycle.org. Accessed 1 Dec 2007

DIAMONDS-D3.5 (2007) Deliverable 3.5 sets of network modules and of cis-regulatory motifs. http://www.sbcellcycle.org. Accessed 1 Dec 2007

Diez et al.(2007) Codelink: an R package for analysis of GE healthcare gene expression bioarrays. Bioinformatics 23:1168–1169

Dowell RD, Jokerst RM, Day A, Eddy SR, Stein L (2001) The distributed annotation system. BMC Bioinformatics 2:7. http://www.biodas.org. Accessed 1 Dec 2007

DNA Repair (2007) DNA damage response and repair mechanisms. http://www.dna-repair.nl. Accessed 1 Dec 2007

DREAM/ENFIN (2007) ENFIN wiki. http://www.enfin.org/dokuwiki/doku.php?id=wiki:wp7. Accessed 1 Dec 2007

DTU-Physics (2007) Technical University of Denmark – Department of Physics. http://www.fys.dtu.dk/English/Research1.aspx. Accessed 1 Dec 2007

EAMNET (2007) European Advanced Light Microscopy Network. http://www.embl-heidelberg.de/eamnet/. Accessed 1 Dec 2007

EAMNET-Teaching (2007) Teaching modules. http://www.embl-heidelberg.de/eamnet/html/teaching_modules.html. Accessed 1 Dec 2007

EB-eye (2007) EBI search tool. http://www.ebi.ac.uk/Information/News/pdf/Press11Dec06.pdf. Accessed 1 Dec 2007

EBI (2007) European Bioinformatics Institute of the European Molecular Biology Laboratory at Hinxton. http://www.ebi.ac.uk/. Accessed 1 Dec 2007

EBI-2can (2007) EBI bioinformatics educational resource. http://www.ebi.ac.uk/2can/home.html. Accessed 1 Dec 2007

EBIMed (2007) Information retrieval and extraction from Medline. http://www.ebi.ac.uk/Rebholz-srv/ebimed/index.jsp. Accessed 1 Dec 2007
eBioSci (2007) A European platform for access and retrieval of full text and factual information in the life sciences. http://www.e-biosci.org/. Accessed 1 Dec 2007
Editorial Nature Genetics (2005) WayStation to HUGOBase. Nat Genet 37(8):783
Editorial Nature Genetics (2006) Jousting for HUGOBase. Nat Genet 38(6):599
EEL (2007) Enhancer element locator. http://www.cs.helsinki.fi/u/kpalin/EEL/. Accessed 1 Dec 2007
EFPIA (2007) European pharmaceutical industry association. http://www.efpia.org. Accessed 1 Dec 2007
EGEE (2007) Enabling grids for e-science in Europe. http://www.eu-egee.org/. Accessed 1 Dec 2007
Epstein RJ (2003) Human molecular biology – an introduction to the molecular basis of health and disease. Cambridge University Press, Cambridge
eHealth (2007) eHealth portfolio of projects for FP6 (2007) http://www.ehealthnews.eu/content/view/753/62/. Accessed 1 Dec 2007
ELIXIR (2007) European life sciences infrastructure for biological information. http://www.elixir-europe.org. Accessed 1 Dec 2007
ELMI (2007) European Light Microscopy Initiative. http://cci.sahlgrenska.gu.se/ELMI/. Accessed 1 Dec 2007
EMBL (2007) European Molecular Biology Laboratory. http://www.embl.org/. Accessed 1 Dec 2007
EMBL-Bank (2007) Nucleotide sequence database. http://www.ebi.ac.uk/embl/. Accessed 30 Aug 2007
EMBL-Hamburg Outstation (2007) EMBL Hamburg Outstation http://www.embl.org/sites/hhsite.html. Accessed 1 Dec 2007
EMBL-Heidelberg (2007) European Molecular Biology Laboratory at Heidelberg. http://www.embl-heidelberg.de/. Accessed 1 Dec 2007
EMBL-Strategic (2007) EMBL 2007 strategic forward look 2006–2015. http://www.embl.org/aboutus/news/publications/pdf/sfl06.pdf. Accessed 1 Dec 2007
EMBNET (2007) European Molecular Biology Network. http://www.embnet.org. Accessed 1 Dec 2007
EMBOSS (2007) The European molecular biology open software suite. http://emboss.org. Accessed 1 Dec 2007
EMBRACE (2007) A European model for bioinformatics research and community education – bioinformatics grid. http://www.embracegrid.info. Accessed 1 Dec 2007
EMBRACE-Biomed (2007) EMBRACE information for biomedical scientists. http://www.embracegrid.info/page.php?page=info_biomed. Accessed 1 Dec 2007
EMBRACE-GUIDE (2007) EMBRACE Web services development guide. http://www.embracegrid.info/page.php?page=tech_documents. Accessed 1 Dec 2007
EMBRACE-Proposal (2007) Project proposal. http://www.embracegrid.info/page.php?page=proposal_txt. Accessed 1 Dec 2007
EMBRACE-WkP4 (2007) Work package 4. https://bioinformatics.bmc.uu.se/WP4/. Accessed 1 Dec 2007
EMI-CD (2007) European modelling initiative combating complex diseases. http://pybios.molgen.mpg.de/EMICD/. Accessed 1 Dec 2007
EMI-CD-APOPTOSIS (2007) A kinetic model for apoptosis. http://pybios.molgen.mpg.de/EMICD/Deliverables/apoptosis.pdf. Accessed 1 Dec 2007
EMMA (2007) European Mouse Mutant Archive. http://www.emma.rm.cnr.it. Accessed 1 Dec 2007
EMSD (2007) The European Molecular Structure Database. http://www.ebi.ac.uk/msd/Temblor/Temblor1.html. Accessed 1 Dec 2007
ENCODE (2007) Encyclopedia of DNA elements. http://www.genome.gov/10005107. Accessed 1 Dec 2007
ENCODE-Participants (2007) (ENCODE 2007 lists of participants and projects). http://www.genome.gov/12513391#1. Accessed 1 Dec 2007

ENCODE-Project-Consortium (2007) Identification and analysis of functional elements in 1% of the human genome by the ENCODE pilot project. Nature 447:799–816. http://www.nature.com/nature/journal/v447/n7146/full/nature05874.html. Accessed 1 Dec 2007

ENFIN (2007) Enabling systems biology PP6 (2007) project. http://www.enfin.org. Accessed 1 Dec 2007

ENFIN-wp5.2 (2007) ENFIN (2007) Work package 5.2 predicting new genes in partially characterized signalling pathways from the combined signal of protein-protein interactions and cis-regulatory regions. http://www.enfin.org/page.php?page=wp&wp=5. Accessed 1 Dec 2007

Ensembl (2007) Automatic annotation on selected eukaryotic genomes. http://www.ensembl.org. Accessed 1 Dec 2007

Entrez (2007) The life sciences search engine. http://www.ncbi.nlm.nih.gov/sites/gquery. Accessed 1 Dec 2007

Entrez-models (2007) Models of Entrez databases. http://www.ncbi.nih.gov/Database/datamodel. Accessed 1 Dec 2007

EPISTEM (2007) Role of p63 and related pathways in epithelial stem cell proliferation and differentiation and in rare ectrodactyly ectodermal dysplasia – related syndromes. http://www.epistem.eu/. Accessed 1 Dec 2007

EPO (2007) European Patent Office. http://www.epo.org/. Accessed 1 Dec 2007

EPO-Patent-Search (2007) EPR 2007 patent search tool. http://www.espacenet.com. Accessed 1 Dec 2007

EPSS (2007) Electronic proposal preparation system. http://cordis.europa.eu/index.cfm?fuseaction=UserSite.FP7SubmitProposalPage. Accessed 1 Dec 2007

ERA (2007) European research area. http://cordis.europa.eu/era/. Accessed 1 Dec 2007

ESB (2007) European School of Bioinformatics. http://www.biosapiens.info/page.php?page=esb. Accessed 1 Dec 2007

ESBIC-D (2007) European systems biology initiative combating complex diseases http://pybios.molgen.mpg.de/ESBIC-D/. Accessed 1 Dec 2007

ESB-Schools (2007a) ESB 2007 schools. http://cassandra.bio.uniroma1.it/biosap/Euroschool1. Accessed 1 Dec 2007

ESB-Schools (2007b) ESB 2007 schools. http://cassandra.bio.uniroma1.it/ESF04, Accessed 1 Dec 2007

ESB-Schools (2007c) ESB 2007 schools. http://www.cmbi.kun.nl/euroschool/index.html. Accessed 1 Dec 2007

ESB-Schools (2007d) ESB 2007 schools. http://www.cmbi.kun.nl/euroschool/membrane.html. Accessed 1 Dec 2007

ESB-Schools (2007e) ESB 2007 schools. http://cassandra.bio.uniroma1.it/biosap/Euroschool3. Accessed 1 Dec 2007

ESB-Schools (2007f) ESB 2007 schools. http://pen2.igc.gulbenkian.pt/4ESB/. Accessed 1 Dec 2007

ESB-Schools (2007g) ESB 2007 schools. http://mispred.enzim.hu/course/index.htm. Accessed 1 Dec 2007

ESF (2007) The European Science Foundation. http://www.esf.org/. Accessed 1 Dec 2007

ESFRI (2007) European Strategy Forum on Research Infrastructures http://cordis.europa.eu/esfri/. Accessed 1 Dec 2007

ESF-SysBio (2007) ESF 2007 forward look on systems biology. http://www.esf.org/activities/forward-looks/life-earth-and-environmental-sciences-lesc/current-forward-looks-in-life-earth-and-environmental-sciences/systems-biology.html. Accessed 1 Dec 2007

ESF-Task-Force (2007) ESF 2007 task force on systems biology – strategic guidance and recommendations. http://www.esf.org/publications.html. Accessed 1 Dec 2007

EUCLOCK (2007) Entrainment of the circadian clock. http://www.euclock.org/. Accessed 1 Dec 2007

euHCVdb (2007) European hepatitis C virus database. http://euhcvdb.ibcp.fr/euHCVdb/. Accessed 1 Dec 2007

EUMORPHIA (2007) Understanding human disease through mouse genetics. http://www.eumorphia.org/. Accessed 1 Dec 2007

Eurofungbase (2007) European fungal genomic database. http://eurofung.net/index.php?option=com_content&task=section&id=3&Itemid=4 and http://www.eurofung.net. Accessed 1 Dec 2007

EUROHEAR (2007) Bringing the genetic basis of deafness to light, an FP6 research project. http://www.eurohear.org/. Accessed 1 Dec 2007

EUROHEAR-newsletter (2007) EUROHEAR statistics for biochips: the small n, large p paradigm. http://www.eurohear.org/pdf/EUROHEAR_newsletter2.pdf. Accessed 1 Dec 2007

Euromyths (2007) Myths about the European Union. http://ec.europa.eu/unitedkingdom/press/euromyths/index_en.htm. Accessed 1 Dec 2007

Europa (2007) Gateway to the European Union. http://europa.eu/. Accessed 1 Dec 2007

Europhysiome (2007) The European Physiome Consortium. http://www.europhysiome.org/. Accessed 1 Dec 2007

EuroStemCell (2007) European Consortium for Stem Cell Research. http://www.eurostemcell.org/. Accessed 1 Dec 2007

EurSysBio (2007) European Systems Biology. http://www.systembiology.net/. Accessed 1 Dec 2007

EuTRACC (2007) European Transcriptome, Regulome, Cellular Commitment Consortium http://www.eutracc.eu/. Accessed 1 Dec 2007

EVI-GENEROT (2007) European Vision Institute – functional genomics of the retina in health and disease. http://www.evi-genoret.org. Accessed 1 Dec 2007

eVOContology (2007) Orthogonal controlled vocabularies that unifies gene expression data. http://www.evocontology.org/. Accessed 1 Dec 2007

Expression-Profiler (2007) An open, extensible web-based collaborative platform for microarray gene expression, sequence and PPI data analysis. http://www.ebi.ac.uk/expressionprofiler/. Accessed 1 Dec 2007

Fall CP, Marland ES, Wagner JM, Tyson JJ (2002) Computational cell biology. Springer, New York

Faure J-E, Marcus F, Sambain B (2005) Future needs for research infrastructures in biomedical sciences in Europe. ftp://ftp.cordis.europa.eu/pub/lifescihealth/docs/bms_infrastructures_workshop_report.pdf. Accessed 1 Dec 2007

FELICS (2007) Free European Life-Science Information and Computational Services. http://www.felics.org. Accessed 1 Dec 2007

FIL (2007) Functional imaging laboratory of the Institute of Neurology, University College London. http://www.fil.ion.ucl.ac.uk. Accessed 1 Dec 2007

FlyBase (2007) A database of drosophila genes and genomes. http://flybase.bio.indiana.edu/. Accessed 1 Dec 2007

Food (2007) Food quality and safety – FP6 major projects library. http://ec.europa.eu/research/fp6/projects.cfm?p=5. Accessed 1 Dec 2007

FP5 (2007) Framework Programme 5: the Fifth Framework Programme 1998–2002. http://cordis.europa.eu/fp5/. Accessed 1 Dec 2007

FP6 (2007) Framework Programme 6: the Sixth Framework Programme 2002–2006. http://ec.europa.eu/research/fp6/index_en.cfm. Accessed 1 Dec 2007

FP6-EoI (2007) Framework Programme 6 expressions-of-interest http://cordis.europa.eu/eoi/search_form.cfm. Accessed 1 Dec 2007

FP6-evaluation (2007) Framework Programme 6 evaluation procedures. http://cordis.europa.eu/fp6/stepbystep/eval.htm. Accessed 1 Dec 2007

FP6-instruments (2007) Framework Programme 6 instruments. http://cordis.europa.eu/fp6/instruments.htm. Accessed 1 Dec 2007

FP6–2002-LIFESCIHEALTH (2002) Thematic call in the area of "life sciences, genomics and biotechnology for health" http://cordis.europa.eu/fp6/dc/index.cfm?fuseaction=UserSite.LifeSciHealthDetailsCallPage&call_id=4. Accessed 1 Dec 2007.

FP6-Projects (2007) Find a project. http://cordis.europa.eu/fp6/projects.htm and http://www.lifecompetence.cu. Accessed 1 Dec 2007

FP6 references (2007) Framework Programme 6 instruments. http://cordis.europa.eu/fp6/projects.htm. Accessed 1 Dec 2007.

FP6-step-by-step (2007) Framework Programme 6 structured walk-through of what FP6 participation entails. http://cordis.europa.eu/fp6/stepbystep/home.html. Accessed 1 Dec 2007

FP7 (2007) Framework Programme 7: the future of European Union research policy. http://ec.europa.eu/research/fp7/index_en.cfm. Accessed 1 Dec 2007

FP7-Beneficiaries (2007) Beneficiaries guide. (ftp://ftp.cordis.europa.eu/pub/fp7/docs/beneficiaries_en.pdf. Accessed 1 Dec 2007

FP7-CALL-HEALTH-2007-A (2007) FP7 2007 first call for proposals in health. http://cordis.europa.eu/fp7/dc/index.cfm?fuseaction=UserSite.CooperationDetailsCallPage&call_id=10. Accessed 1 Dec 2007

FP7-CALL-HEALTH-2007-B (2007) FP7 2007 second call for proposals in Health. http://cordis.europa.eu/fp7/dc/index.cfm?fuseaction=UserSite.CooperationDetailsCallPage&call_id=63. Accessed 1 Dec 2007

FP7-Calls-Health (2007) FP7 calls for proposals in health research. http://cordis.europa.eu/fp7/dc/index.cfm?fuseaction=UserSite.CooperationCallsPage&id_activity=1. Accessed 1 Dec 2007

FP7-Checklist (2007) Checklist for consortium agreements ftp://ftp.cordis.europa.eu/pub/fp7/docs/checklist_en.pdf. Accessed 1 Dec 2007

FP7-CORDIS (2007) CORDIS Seventh Framework Research Programme. http://cordis.europa.eu/fp7/home_en.html. Accessed 1 Dec 2007

FP7-Enquiry (2007)Enquiry service. http://ec.europa.eu/research/enquiries. Accessed 1 Dec 2007

FP7-ethics (2007) Ethics. http://cordis.europa.eu/fp7/ethics_en.html#ethics_sd. Accessed 1 Dec 2007

FP7-Financial (2007) Financial guidelines. ftp://ftp.cordis.europa.eu/pub/fp7/docs/financialguide_en.pdf. Accessed 1 Dec 2007

FP7-Find-a-call (2007) Find a call. http://cordis.europa.eu/fp7/dc/index.cfm. Accessed 1 Dec 2007

FP7-Find-Document (2007) Find a document. http://cordis.europa.eu/fp7/find-doc_en.html. Accessed 1 Dec 2007

FP7-Get-Support (2007) Get support. http://cordis.europa.eu/fp7/get-support_en.html. Accessed 1 Dec 2007

FP7-ICT-2007–2 (2007) Information and communications technologies call for proposals. http://cordis.europa.eu/fp7/dc/index.cfm?fuseaction=UserSite.FP7DetailsCallPage&Call_ID=65. Accessed 1 Dec 2007

FP7-Intellectual Property Rights (2007) Intellectual property rights. ftp://ftp.cordis.europa.eu/pub/fp7/docs/ipr_en.pdf. Accessed 1 Dec 2007

FP7-Model-Grant (2007) Model grant agreement. http://cordis.europa.eu/fp7/calls-grant-agreement_en.html#standard_ga. Accessed 1 Dec 2007

FP7-Negotiating (2007) Negotiation guidelines. ftp://ftp.cordis.europa.eu/pub/fp7/docs/negotiation_en.pdf. Accessed 1 Dec 2007

FP7-Other-Support (2007) Other support. http://cordis.europa.eu/fp7/othersupport_en.html. Accessed 1 Dec 2007

FP7-Participate (2007) Participate in FP7. http://cordis.europa.eu/fp7/participate_en.html. Accessed 1 Dec 2007

FP7-Partners (2007)Partners service. http://cordis.europa.eu/fp7/partners_en.html. Accessed 1 Dec 2007

FP7-Proposal-Preparation (2007) Proposal preparation. http://cordis.europa.eu/fp7/dc/index.cfm?fuseaction=UserSite.FP7SubmitProposalPage. Accessed 1 Dec 2007

FP7-Rules (2007) FP7 Rules for participation. http://cordis.europa.eu/documents/documentlibrary/90798691EN6.pdf. Accessed 1 Dec 2007

FP7-Specific-Programme (2007) Council decision 2006/971/EC of 19 December 2006 concerning the specific programme 'Cooperation' implementing the Seventh Framework Programme of the European Community for research, technological development and demonstration activities (2007 to 2013) (OJ L 400, 30.12.2006). http://eur-lex.europa.eu/LexUriServ/site/en/oj/2006/l_400/l_40020061230en00860242.pdf. Accessed 1 Dec 2007

Fundamental-Genomics (2007) Fundamental knowledge and basic tools for functional genomics in all organisms. http://cordis.europa.eu/lifescihealth/genomics/home.htm. Accessed 1 Dec 2007

GAIN (2007) Genetic Association Information Network. http://www.fnih.org/GAIN2/home_new.shtml. Accessed 1 Dec 2007

GALGO (2007) Genetic algorithm variable selection strategy. http://www.bip.bham.ac.uk/bioinf/software.html. Accessed 1 Dec 2007

Gautheret D (2007) Transcript isoform entropy: what is the extent of splicing disruption in cancer. EBI Seminar, 11 June 2007, Hinxton, UK

GEANT2 (2007) Pan-European research and education network. http://www.geant2.net. Accessed 1 Dec 2007

GENCODE (2007) BioSapiens identification of all protein-coding genes in the ENCODE selected regions http://www.pdg.cnb.uam.es/mtress/ENCODE/index.html. Accessed 1 Dec 2007

Gene3D (2007) Domain architecture classification databases. http://gene3d.biochem.ucl.ac.uk/Gene3D/. Accessed 1 Dec 2007

GenomEUtwin (2007) Studies of European volunteer twins to identify genes underlying common diseases. http://www.genomeutwin.org/index.htm. Accessed 1 Dec 2007

Genome Reviews (2007) Genome reviews database. http://www.ebi.ac.uk/GenomeReviews/. Accessed 1 Dec 2007

Genomics-Systems-Biology (2007) The PP7 2007 unit for genomics and systems Biology. http://cordis.europa.eu/fp7/cooperation/health_en.html. Accessed 1 Dec 2007

Geno2pheno (2007) Interpreting genotypic HIV drug resistance tests http://www.geno2pheno.org. Accessed 1 Dec 2007

Geva-Zatorsky N, Rosenfeld N, Itzkovitz S, Milo R, Sigal A, Dekel E, Yarnitzky T, Liron Y, Polak P, Lahav G, Alon U (2006) Oscillations and variability in the p53 system. Mol Syst Biol 2:2006. 0033:1–13

Ghalouci R (2007) Combating deadly diseases – European Union funded projects, 3rd edn. Report EUR 22455, European Commission, Brussels, Belgium

Glue-Grants (2007) NIGMS Glue Grants. http://www.nigms.nih.gov/Initiatives/Collaborative/GlueGrants/. Accessed 1 Dec 2007

GO (2007) Gene Ontology. http://www.ebi.ac.uk/GO/. Accessed 1 Dec 2007

GOA (2007) Gene Ontology Annotation database. http://www.ebi.ac.uk/GOA. Accessed 1 Dec 2007

Granstrand O (2007) Intellectual property rights aspects of Internet collaborations. http://ec.europa.eu/research/era/pdf/ipr-internetbasedresearch-workshopreport.pdf. Accessed 1 Dec 2007

Griffiths AJF, Miller JH, Suzuki DT, Lewontin RC, Gelbart WM (2000) An introduction to genetic analysis, 7th edn. Freeman, New York

GSCAN (2007) The genetic architecture of complex traits in heterogeneous stock mice. http://gscan.well.ox.ac.uk/. Accessed 1 Dec 2007

GSCANDB (2007) GSCAN (2007) Database. http://www.well.ox.ac.uk/rmott/GSCANDB/. Accessed 1 Dec 2007

Gyorffi M, Marcus F (eds) (2003) Bioinformatics – structures for the future. ftp://ftp.cordis.europa.eu/pub/lifescihealth/docs/bioinf_workshoprpt_2003_06_30_final.pdf. Accessed 1 Dec 2007

Hahn WC, Weinberg RA (2002) Modelling the molecular circuity of cancer. Nat Rev Cancer 2(5):331–341

Hainaut P, Wiman KG (eds) (2007) 25 years of p53 research. Springer, New York

Hallikas O, Palin K, Sinjushina N, Rautiainen R, Partanen J, Ukkonen E, Taipale J (2006) Genome-wide prediction of mammalian enhancers based on analysis of transcription-factor binding affinity. Cell 124:47–59

Hanahan D, Weinberg RA (2000) The hallmarks of cancer. Cell 100(1):57–70

HAVANA (2007) Human and vertebrate analysis and annotation. http://www.sanger.ac.uk/HGP/havana. Accessed 1 Dec 2007

HCYCLEP (2007) Human cell CYCLE periodically regulated genes based on protein features. http://www.cbs.dtu.dk/services/hcyclep. Accessed 1 Dec 2007

Health-Research (2007) The FP7 2007 Health Research Directorate. http://cordis.europa.eu/fp7/cooperation/health_en.html. Accessed 1 Dec 2007

HepatoSys (2007) German network systems biology hepatocyte programme. http://www.systembiologie.de/de/index.html; http://www.systembiologie.de/doc/070416Milestones HepatoSysII_Text.pdf. Accessed 1 Dec 2007

Hescheler J, Sachinidis A, Chen S, Winkler J (2006) Functional genomics in engineered embryonic stem cells. Stem Cell Rev 2(1):1–4

HGMD (2007) Human gene mutation database. http://www.hgmd.cf.ac.uk/ac/index.php. Accessed 1 Dec 2007

HGVbase (2007) Human genome variation database. http://hgvbase.org. Accessed 1 Dec 2007

Higgins D, Taylor W (2000) Bioinformatics: sequence, structure and databanks. Oxford University Press, Oxford

Hinsby AM, Kiemer L, Karlberg EO, Hansen KL, Fausbøll A, Juncker AS, Andersen JS, Mann M, Brunak S (2006) A wiring of the human nucleolus. Mol Cell 22(2):285–295

Hoffmann NR et al (2005) Text mining for metabolic pathways, signalling cascades, and protein networks. Science Signal Transduction Knowledge Environment 10 May 2005(283):pe21. Review

Hohmann (2007) Hohmann laboratory – Gothenberg university. http://www.gmm.gu.se/groups/hohmann/. Accessed 1 Dec 2007

Hubbard T, Andrews D, Caccamo M, Cameron G, Chen Y, Clamp M, Clarke L, Coates G, Cox T, Cunningham F, Curwen V, Cutts T, Down T, Durbin R, Fernandez-Suarez XM, Gilbert J, Hammond M, Herrero J, Hotz H, Howe K, Iyer V, Jekosch K, Kahari A, Kasprzyk A, Keefe D, Keenan S, Kokocinsci F, London D, Longden I, McVicker G, Melsopp C, Meidl P, Potter S, Proctor G, Rae M, Rios D, Schuster M, Searle S, Severin J, Slater G, Smedley D, Smith J, Spooner W, Stabenau A, Stalker J, Storey R, Trevanion S, Ureta-Vidal A, Vogel J, White S, Woodwark C, Birney E (2005) Ensembl 2005. Nucleic Acids Res 33:D447–453.

IARC-tp53 (2007) The IARC TP53 mutation database. http://www-p53.iarc.fr/. Accessed 1 Dec 2007

ICBP (2007) Integrative cancer biology programme of the NCI. http://icbp.nci.nihgov. cited 1 Dec 2007.

ICD-9-CM (2007) International classification of diseases, ninth revision, clinical modification. http://www.cdc.gov/nchs/about/otheract/icd9/abticd9.htm. Accessed 1 Dec 2007

iGEM (2007) International genetically engineered machine competition. http://parts.mit.edu/r/parts/igem/index.cgi. Accessed 1 Dec 2007

IGLO (2007) Informal Group of RTD Liaison Offices. http://www.iglortd.org/. Accessed 1 Dec 2007

iHop (2007) Information hyperlinked over proteins. http://www.ihop-net.org/UniPub/iHOP/. Accessed 1 Dec 2007

IIMS (2007) Integrating the results of 3-D electron microscopy. http://www.ebi.ac.uk/msd/projects/IIMS.html. Accessed 1 Dec 2007

IMI (2007) Innovative Medicines Initiative. http://www.imi-europe.org/. Accessed 1 Dec 2007

IMI-Research-Agenda (2007) IMI research agenda. http://www.imi-europe.org/sitecollection/Innovative%20Medecines%20Initiative%20SRA%20version%202.0pdf. Accessed 1 Dec 2007

ImmunoGrid (2007) The European virtual human immune system project http://www.immunogrid.org/. Accessed 1 Dec 2007

INCA (2007) Research on the role of chronic infections in the development of cancer. http://www.inca-project.org. Accessed 1 Dec 2007

INFOBIOMED (2007) Structuring European biomedical informatics to support individualised healthcare. http://www.infobiomed.org/. Accessed 1 Dec 2007

INFOBIOMED-Wiki (2007) Wiki of INFOBIOMED. http://139.91.190.38/InfobiomedWiki/index.php. Accessed 1 Dec 2007

Infosoc (2007) The Directorate General for the Information Society of the European Commission – information and communication technologies. http://cordis.europa.eu/fp7/ict/. Accessed 1 Dec 2007

Ingemansson T, Knezevic M (2005) 100 Technology offers stemming from EU biotechnology RTD results. European Commission report EUR 20603. ftp://ftp.cordis.europa.eu/pub/lifescihealth/docs/booklet_100_off.pdf. Accessed 1 Dec 2007

Innomed (2007) Innovative medicines. http://www.nsmf.org/Attachments/innomed.htm. Accessed 1 Dec 2007

INRA (2007) Institut National de la Recherche Agronomique. http://www.basysbio.eu/institutions.php?id=1. Accessed 1 Dec 2007

INSERM (2007) INSERM advanced technologies for genomics and clinical research. http://www.mayeticvillage.com/QuickPlace/it-atdproject/Main.nsf/h_Index/FAEF77AD27F93F4BC1256FBE0035D515/?OpenDocument. Accessed 1 Dec 2007

IntAct (2007) Protein interaction database and tools. http://www.ebi.ac.uk/intact. Accessed 1 Dec 2007

Integr8 (2007) Access to complete genomes and proteomes. http://www.ebi.ac.uk/integr8. Accessed 1 Dec 2007

InterPro (2007) A database of protein families, domains and functional sites. http://www.ebi.ac.uk/interpro/. Accessed 1 Dec 2007

IPI (2007) International Protein Index. http://www.ebi.ac.uk/IPI/IPIhelp.html. Accessed 1 Dec 2007

IRC (2007) The International Regulome Consortium. http://www.internationalregulomeconsortium.ca/. Accessed 1 Dec 2007

Janson M, Loughlin R, Loiodice I, Fu C, Brunner D, Nédélec F, Tran P (2007) Crosslinkers and motors organize dynamic microtubules to form stable bipolar arrays in fission yeast. Cell 128:357–368

Jehensen P, Marcus F (eds) (2005) EU Projects Workshop Report on systems biology. IEE Proc Syst Biol 152(2):55–60. ftp://ftp.cordis.europa.eu/pub/lifescihealth/docs/systems_biology_worskhop_report_jan2005.pdf. Accessed 1 Dec 2007

Jobling MA, Hurles ME, Tyler-Smith C (2004) Human evolutionary genetics – origins, peoples and disease. Garland, London

Johnston MD, Edwards CM, Bodmer WF, Maini PK, Chapman SJ (2007) Mathematical modeling of cell population dynamics in the colonic crypt and in colorectal cancer. Proc Natl Acad Sci USA 104(10):40084013.http://www.pnas.org/cgi/reprint/104/10/4008.pdf. Accessed 1 Dec 2007

Joliff-Botrel G, Perrin (2007) Stem cells – European research projects involving stem cells in the 6th Framework Programme. ftp://ftp.cordis.europa.eu/pub/fp7/docs/stemcell_eu_research_fp6_en.pdf. Accessed 1 Dec 2007

Kaku M (1998) Visions – how science will revolutionize the twenty-first century. Oxford University Press, Oxford

Kapushesky M, Kemmeren P, Culhane AC, Durinck S, Ihmels J, Krner C, Kull M, Torrente A, Sarkans U, Vilo J, Brazma A (2004) Expression profiler: next generation-an online platform for analysis of microarray data. Nucleic Acids Res 32:W465–W470

KEGG (2007) Kyoto encyclopaedia of genes and genomes. http://www.genome.jp/kegg/. Accessed 1 Dec 2007

Kersey P, Bower L, Morris L, Horne A, Petryszak R, Kanz C, Kanapin A, Das U, Michoud K, Phan I, Gattiker A, Kulikova T, Faruque N, Duggan K, Mclaren P, Reimholz B, Duret L, Penel S, Reuter I, Apweiler R (2005) Integr8 and genome reviews: integrated views of complete genomes and proteomes. Nucleic Acids Res 33:D297D302. http://nar.oxfordjournals.org/cgi/content/full/33/suppl_1/D297. Accessed 1 Dec 2007

Kim D, Rath O, Kolch W, Cho KH (2007) A hidden oncogenic positive feedback loop caused by crosstalk between Wnt and ERK pathways. Oncogene 26(31):4571–4579

Kinetikon (2007) Biochemical reaction kinetics database. http://kinetikon.molgen.mpg.de/. Accessed 1 Dec 2007

Kitano H (ed) (2001) Foundations of systems biology. MIT Press, Cambridge

Klefenz H (2002) Industrial pharmaceutical biotechnology. Wiley-VCH, Weinheim

Klipp E, Herwig Kowald A, Wierling C, Lehrach H (2005) Systems biology in practice – concepts, implementation and application. Wiley-VCH, Weinheim

Klipp E, Liebermeister W, Helbig A, Kowald A, Schaber J (2007) Standards in computational systems biology. http://www.ebi.ac.uk/biomodels/doc/KlippSurvey.pdf. Accessed 1 Dec 2007

Koch et al.(2007) The landscape of histone modifications across 1% of the human genome in five human cell lines. Genome Res 17:691–707

Krebs HA (1953) 1953 Nobel prize in medicine or physiology. http://nobelprize.org/nobel_prizes/medicine/laureates/1953/index.html. Accessed 1 Dec 2007

Krishna R, Schaefer HG, Bjerrum OJ (2007) Effective integration of systems biology, biomarkers, biosimulation and modelling in streamlining drug development. Eur J Pharm Sci 31(1):62–67

Krull M, Pistor S, Voss N, Kel A, Reuter I, Kroneberg D, Michael H, Schwarzer K, Potapov A, Choi C, Kel-Margoulis O, Wingender E (2006) TRANSPATH®: an information resource for storing and visualizing signalling pathways and their pathological aberrations. Nucleic Acids Res 34:D546–D551

Kumar P, Clark M (2002) Clinical medicine, 5th edn. Saunders, Edinburgh

Kyriakopoulou C et al (eds) (2007) From fundamental genomics to systems biology – project catalogue. http://cordis.europa.eu/fp7/health/library_en.html. Accessed 1 Dec. 2007

Lage K et al.(2007) A human phenome-interactome network of protein complexes implicated in genetic disorders. Nat Biotechnol 25(3):309316.doi:10.1038/nbt1295

Laycock J, Wise P (1996) Essential endocrinology, 3rd edn. Oxford University Press, Oxford

Lengauer T (ed) (2007) Bioinformatics – from genomes to therapies. Wiley-VCH, Weinheim

Lesk AM (2001) Introduction to protein architecture. Oxford University Press, Oxford

Lesk AM (2002) Introduction to bioinformatics. Oxford University Press, Oxford

Le Texier V, Le Riethoven J-J, Kumanduri V, Gopalakrishnan C, Lopez F, Gautheret D, and Thanaraj TA (2006) AlTrans: transcript pattern variants annotated for both alternative splicing and alternative polyadenylation. BMC Bioinformatics 7:169

Lichtenberg U, Jensen S, Jensen L, Brunak S (2003) Protein feature based identification of cell cycle regulated proteins in yeast. J Mol Biol 329:663–674

LMU (2007) Ludwig-Maximilians-University, Munich – Institute for Medical Psychology, Centre for Chronobiology. http://www.uni-muenchen.de. Accessed 1 Dec 2007

LOVD (2007) Leiden Open Variation Database. http://www.dmd.nl/LOVD. Accessed 1 Dec 2007

Love CA et al.(2003) The ligand-binding face of the semaphorins revealed by the high-resolution crystal structure of SEMA4D. Nat Struct Biol 10(10):843–848

LUIB (2007) Leiden University Institute of Biology, section Microbiology. http://www.leiden.edu/. Accessed 1 Dec 2007

Lymphangiogenomics (2007) Genome-wide discovery and functional analysis of novel genes in lymphangiogenesis. http://www.lymphomic.org/. Accessed 1 Dec 2007

Macdonald F, Ford CHJ, Casson AG (2004) Molecular biology of cancer, 2BIOS Scientific, Londonnd edn.

MAGE (2007) MicroArray and gene expression. http://www.mged.org/Workgroups/MAGE/mage.html. Accessed 1 Dec 2007

Manoussaki E (ed) (2006) Cancer research – projects funded under the Sixth Framework Programme. Report EUR 22051, European Commission. ftp://ftp.cordis.europa.eu/pub/lifescihealth/docs/general_catalogue_en.pdf

Mantela J, Jiang Z, Ylikoski J, Fritzsch B, Zacksenhaus E, Pirvola U (2005) The retinoblastoma gene pathway regulates the postmitotic state of hair cells of the mouse inner ear. Development 132:2377–2388

Marcus F, Mulligan B, Sansom M (2004) Computational systems biology (CSB) – its future in Europe ftp://ftp.cordis.europa.eu/pub/lifescihealth/docs/csbworkshop_2004_03_en.pdf. Accessed 1 Dec 2007

Marcus F, Mulligan B (2006) Workshop on European database and analysis resources for research in human genetic variation. ftp://ftp.cordis.europa.eu/pub/lifescihealth/docs/geneticvariationworkshop-finalreport_200604.pdf. Accessed 1 Dec 2007

MGED (2007) Ontology for describing microarray experiments. http://mged.sourceforge.net/ontologies/index.php. Accessed 1 Dec 2007

MIAME (2007) Minimum Information About a Microarray Experiment. http://www.mged.org/Workgroups/MIAME/miame.html. Accessed 1 Dec 2007

MIAMExpress (2007) a MIAME compliant microarray data submission tool. http://www.ebi.ac.uk/miamexpress/. Accessed 1 Dec 2007

MiMage (2007) Role of mitochondria in conserved mechanisms of ageing. http://www.mimage.uni-frankfurt.de/. Accessed 1 Dec 2007

MIT-Open-Courseware (2007) MIT free and open educational resource http://ocw.mit.edu/index.html. Accessed 1 Dec 2007

MitoCheck (2007) Regulation of mitosis by phosphorylation – a combined functional genomics, proteomics and chemical biology approach. http://www.mitocheck.org/. Accessed 1 Dec 2007

MolPage (2007) Molecular phenotyping to accelerate genomic epidemiology. http://www.molpage.org/. Accessed 1 Dec 2007

MolTools (2007) Advanced molecular tools for array-based analyses of genomes. http://www.moltools.org/. Cited 1 Dec 2007

Mount DW (2001) Bioinformatics: sequence and genome analysis. Cold Spring Harbor Laboratory Press, Cold Spring Harbor

MPI-INF-Bioinformatics-for-HIV (2007) Max Planck Institutes – informatics –bioinformatics for HIV. http://www.mpi-inf.mpg.de/departments/d3/areas/hiv.html. Accessed 1 Dec 2007

MPIMG (2007) Max Planck Institute for Molecular Genetics – Department Lehrach vertebrate genomics. http://www.molgen.mpg.de/research/lehrach/. Accessed 1 Dec 2007

MSD (2007) The EBI 2007 Macromolecular Structure Database. http://www.ebi.ac.uk/msd. Accessed 1 Dec 2007

Muldur U, Corvers F, Delanghe H, Dratwa J, Heimberger D, Sloan B, Vanslembrouck S (2006) A new deal for an effective European research policy – the design and impacts of the 7th Framework Programme. Springer, Berlin, Germany

Mutp53 (2007) Mutant p53 as a target for cancer treatment. http://www.mutp53.com. Accessed 1 Dec 2007

Nagl S (ed) (2006) Cancer bioinformatics: from therapy design to treatment. Wiley, Chichester

Nano2Life (2007) Bringing nanotechnologies to life. http://www.nano2life.org. Accessed 1 Dec 2007

NASC (2007) The European Arabidopsis Stock Centre. http://arabidopsis.info/. Accessed 1 Dec 2007

NCBI (2007) USA National Center for Biotechnology Information of the NIH. http://www.ncbi.nlm.nih.gov/. Accessed 1 Dec 2007

NCBI-Education (2007) NCBI education http://www.ncbi.nlm.nih.gov/Education/index.html. Accessed 1 Dec 2007

NCI (2007) National Cancer Institute of the NIH. http://www.cancer.gov/. Accessed 1 Dec 2007

NCPs (2007) Network of national contact points in member states and associated states. http://cordis.europa.eu/fp7/ncp_en.html. Accessed 1 Dec 2007

NEWT (2007) The Uniprot taxonomy browser. http://www.ebi.ac.uk/newt/display. Accessed 1 Dec 2007

NHGRI (2007) National Human Genome Research Institute of the NIH. http://www.genome.gov/. Accessed 1 Dec 2007

NIGMS (2007) The National Institute of General Medical Sciences. http://www.nigms.nih.gov/. Accessed 1 Dec 2007

NIH (2007) National Institutes of Health of the USA – United States of America. http://www.nih.gov/. Accessed 1 Dec 2007

Noble D (2002) Modelling the Heart-from Genes to Cells to the whole Organ. Science 295 (5560):1678–1682

Noble D (2007) The music of life – biology beyond the genome. Oxford University Press, Oxford

NUCLEOLUS (2007) A wiring of the human nucleolus. http://www.cbs.dtu.dk/suppl/nucleolus/. Accessed 1 Dec 2007

Nugene (2007) Northwestern university gene project. http://www.nugene.org/. Accessed 1 Dec 2007

NuReBase (2007) A reference database on nuclear hormone receptors. http://www.ens-lyon.fr/LBMC/laudet/nurebase/nurebase.html. Accessed 1 Dec 2007

OBO (2007) Open Biomedical Ontologies. http://obofoundry.org/. Accessed 1 Dec 2007

OCISB (2007) Oxford Centre for Integrative Systems Biology. http://www.bioch.ox.ac.uk/sysbio/. Accessed 1 Dec 2007

Official-Journal (2007) Official Journal of the European Union. http://eur-lex.europa.eu/JOIndex.do. Accessed 1 Dec 2007

OMG (2007) The Object Management Group. http://www.omg.com/. Accessed 1 Dec 2007

OMIM (2007) Online Mendelian inheritance in man. http://www.ncbi.nlm.nih.gov/Omim. Accessed 1 Dec 2007

ORegAnno (2007) Open regulatory annotation database. http://www.oreganno.org/oregano/Index.jsp. Accessed 1 Dec 2007

Orengo CA, Thornton JM, Jones DT (2002) Bioinformatics. Bios Scientific, Oxford

Pagana KD, Pagana TJ (2002) Mosby's manual of diagnostic and laboratory tests, 2Mosby, Londonnd edn.

Patrinos and Brookes (2005) DNA, diseases and databases: disastrously deficient. Trends Genet 21:333

PDB (2007) Protein Data Bank. http://www.rcsb.org/pdb. Accessed 1 Dec 2007

Perry JJ, Staley JT, Lory S (2002) Microbial life. Sinauer, Sunderland

Petitjean A, Mathe E, Kato S, Ishioka C, Tavtigian SV, Hainaut P, Olivier M (2007) Impact of mutant p53 functional properties on TP53 mutation patterns and tumour phenotype: lessons from recent developments in the IARC TP53 database. Hum Mutat 28(6):622–629

Pevsner J (2003) Bioinformatics and functional genomics. Wiley-Liss, Hoboken

Physiome (2007) The IUPS physiome project. http://www.physiome.org. Accessed 1 Dec 2007

Polanski A, Kimmel M (2007) Bioinformatics. Springer, New York

PONGO (2007) A web server for multiple predictions of all-alpha transmembrane proteins. http://pongo.biocomp.unibo.it. Accessed 1 Dec 2007

PRIME (2007) Priorities for mouse functional genomic research across Europe. http://www.prime-eu.org/euromouseiiprojects.htm. Accessed 1 Dec 2007

ProDom (2007) Protein domain families database. http://prodom.prabi.fr/prodom/current/html/home.php. Accessed 1 Dec 2007

PROSITE (2007) Database of protein domains, families and functional sites. http://www.expasy.org/prosite/. Accessed 1 Dec 2007

PSB (2007) Plant systems biology – University of Gent. http://www.psb.ugent.be/. Accessed 1 Dec 2007

PSB-UGENT-Software (2007) Bioinformatics and evolutionary genomics software. http://bioinformatics.psb.ugent.be/software. Accessed 1 Dec 2007

PubMed (2007) Citations from biomedical literature. http://www.ncbi.nlm.nih.gov/sites/entrez?db=pubmed. Accessed 1 Dec 2007

PyBioS (2007) A tool for modeling and simulation of cellular systems. http://pybios.molgen.mpg.de/. Accessed 1 Dec 2007

QUASI (2007) Quantifying signal transduction. http://www.idp.mdh.se/quasi. Accessed 1 Dec 2007

Rang HP, Dale MM, Ritter JM (2002) Pharmacology, 4th edn. ChurchillLivingstone, Edinburgh

Reactome (2007) A curated knowledgebase of biological pathways. http://www.reactome.org. Accessed 1 Dec 2007

Reactome-1756 (2007) A curated knowledgebase of biological pathways – path 1756 – phosphorylation of p53 at ser-15 by ATM kinase [Homo sapiens]. http://www.reactome.org/cgi-bin/eventbrowser_st_id?ST_ID=REACT_1756. Accessed 1 Dec 2007

Rebollo E, Sampaio P, Januschke J, Varmark H, Llamazares S, Gonzalez C (2007) Functionally unequal centrosomes drive spindle orientation in asymmetrically dividing drosophila neural stem cells. Dev Cell Mar; 12(3):467–474

Rega (2007) Rega Institute, Katholieke Universiteit Leuven. http://www.kuleuven.be/rega/. Accessed 1 Dec 2007

Regulation (2006) Regulation (EC) no 1906/2006 of the European Parliament and of the Council of 18 December 2006 laying down the rules for the participation of undertakings, research centres and universities in actions under the seventh framework programme and for the dissemination of research results (2007–2013). Off J Eur Union L391/1–L391/18

REGULATORY-GENOMICS (2007) Regulatory genomics FP6 research project. http://research.med.helsinki.fi/regulatorygenomics/. Accessed 1 Dec 2007

RegulonDB (2007) Mechanisms of transcriptional regulation. http://regulondb.ccg.unam.mx/index.html. Accessed 1 Dec 2007

Research-Infrastructures (2007) Research infrastrucures projects in FP6. http://cordis.europa.eu/infrastructures/projects.htm. Accessed 1 Dec 2007

RIBOSYS (2007) Systems biology of RNA metabolism in yeast. http://www.ribosys.org/. Accessed 1 Dec 2007

Roitt I, Brostoff J, Male D (2001) Immunology, 6th edn. Mosby, St Louis

RSAT (2007) Regulatory sequence analysis tools. http://rsat.scmbb.ulb.ac.be/rsat/. Accessed 1 Dec 2007

Sanga S, Sinak J, Frieboes B, Ferrari M, Freuhauf J, Cristini V (2006) Mathematical modeling of cancer progression and response to chemotherapy. Expert Rev Anticancer Ther 6(10):1361–1376

SBI (2007) Systems biology and informatics – University of Rostock. www.sbi.uni-rostock.de. Accessed 1 Dec 2007

SBML (2007) Systems Biology Markup Language. http://sbml.org. Accessed 1 Dec 2007

Schmaltz C (2007) Influenza research – EU funded projects 2001–2007. http://ec.europa.eu/research/health/poverty-diseases/doc/influenza-research_en.pdf. Accessed 1 Dec 2007

Sensen CW (2006) Essentials of genomics and bioinformatics. Wiley, Weinheim

Shannon MT, Wilson BA, Stang CL (2004) Health professional's drug guide. Pearson/Prentice-Hall, New York

Shay JW, Roninson IB (2004) Hallmarks of senescence in carcinogenesis and cancer therapy. Oncogene 23(16):2919–2933

Silver L (2007) The year of miracles. Newsweek 15 Oct 38–43

Singleton PS (2004) Bacteria in biology, biotechnology and medicine, 6th edn. Wiley, Chichester

SmartCell (2007) A cell network simulation program. http://smartcell.embl.de/about.html. Accessed 1 Dec 2007

Smith HJ (2006) Smith and Williams' introduction to the principles of drug design and action, 4th edn. Taylor & Francis, Boca Raton, USA

Sneader W (2005) Drug discovery – a history. Wiley, New York

SNOMED (2007) Systematized nomenclature of medicine-clinical terms. http://www.ihtsdo.org/. Accessed 1 Dec 2007

SPICE (2007) A browser for protein sequences, structures and their annotations. http://www.efamily.org.uk/software/dasclients/spice/index.shtml. Accessed 1 Dec 2007

SPINE (2007) Structural proteomics in Europe. http://www.spineurope.org. Accessed 1 Dec 2007

SRS (2007) Sequence retrieval system. http://srs.ebi.ac.uk. Accessed 1 Dec 2007

Stamm S, Riethoven J-J, Texier V, Le Gopalakrishnan C, Kumanduri V, Tang Y, Barbosa-Morais NL, Thanaraj TA (2006) ASD: a bioinformatics resource on alternative splicing. Nucleic Acids Res 34:D46–D55

Stelling J (2004) Systems analysis of robustness in cellular networks. Shaker, Aachen

Stockwell GR, Thornton JM (2006) Conformational diversity of ligands bound to proteins. J Molecular Biology 356(4):928–944

Strachan T, Read AP (2004) Human molecular genetics, 3rd edn. Garland, New York

STREPTOMICS (2007) Systems biology strategies and metabolome engineering for the enhanced production of recombinant proteins in Streptomyces. http://www.streptomics.org/. Accessed 1 Dec 2007

STRING (2007) Search tool for the retrieval of interacting proteins. http://string.embl.de. Accessed 1 Dec 2007

Swiss-Prot (2007) Protein knowledgebase. http://expasy.org/sprot/. Accessed 1 Dec 2007

SYMBIONIC (2007) Systems biology of a neuronal cell. http://www.symbionicproject.org. Accessed 1 Dec 2007

Symbionic-Workshop (2005) Symbionic workshop on neurogenomics. http://www.symbionicproject.org/Categories.asp?article_category_id_start=4&article_category_id=41. Accessed 1 Dec 2007

Synthetic-Biology (2007) Synthetic biology – applying engineering to biology. Report of a NEST high level group. ftp://ftp.cordis.europa.eu/pub/nest/docs/syntheticbiology_b5_eur21796_en.pdf. Accessed 1 Dec 2007

Sysbiomodels (2007) A list of model repositories. http://www.systems-biology.org/001. Accessed 1 Dec 2007

SYSBIOMED (2007) Systems biology for medical applications. http://www.sysbiomed.org. Accessed 1 Dec 2007

SysMO (2007) Systems biology of micro organisms. http://www.sysmo.net/ and http://www.fz-juelich.de/ptj/datapool/page/2330/European%20SysMO%20concept%20final.pdf. Accessed 1 Dec 2007

SysProt (2007) The integration of proteomics data into systems biology. http://www.sysprot.eu/. Accessed 1 Dec 2007

T-Reg (2007) Relational database on transcriptional regulation. http://treg.molgen.mpg.de/cgi-bin/pfm_meme_form.pl. Accessed 1 Dec 2007

TACB (2005) Therapeutic applications of computational biology and chemistry 2005 conference. http://embl.org.embl.de/aboutus/news/publications/newsletter/issue29.pdf. Accessed 1 Dec 2007

TACB (2007) Therapeutic applications of computational biology and chemistry 2007 conference. http://www.ebi.ac.uk/Information/events/therapeutic/. Accessed 1 Dec 2007

TAVERNA (2007) Workflow and distributed compute technology. http://taverna.sourceforge.net/. Accessed 1 Dec 2007

TCGA (2007). The Cancer Genome Atlas, of the NCI and NHGRI. http://cancergenome.nih.gov. Accessed 1 Dec 2007

TEMBLOR (2007) The European molecular biology linked original resources. http://www.ebi.ac.uk/Information/funding/temblor.html. Accessed 1 Dec 2007

Thanaraj TA, Stamm S, Clark F, Riethoven JJM, Texier V, Le and Muilu J (2004) ASD: the alternative splicing database. Nucl Acids Res 32:D64–D69

The International Human Genome Mapping Consortium (2001) A physical map of the human genome. Nature 409:934–941

Thiery JP, Sleeman JP (2006) Complex networks orchestrate epithelial-mesenchymal transitions. Nat Rev Mol Cell Biol 7:131–142

Tiffin N, Kelso JF, Powell AR, Pan H, Bajic VB, Hide WA (2005) Integration of text- and data-mining using ontologies successfully selects disease gene candidates. Nucleic Acids Res 33(5):1544–1552

TRAMPLE (2007) Transmembrane protein labelling environment. http://www.biocomp.unibo.it/Biosapiens/t.html. Accessed 1 Dec 2007

TRANSFAC (2007) Eukaryotic transcription factors database. http://www.gene-regulation.com/pub/databases.html#transfac. Accessed 1 Dec 2007

TRANSPATH (2007) Signalling pathways and their pathological aberrations. http://www.biobase-international.com/pages/index.php?id=transpathdatabases. Accessed 1 Dec 2007

TreeDet (2007) Tree determinant server. http://www.pdg.cnb.uam.es/Servers/treedet/. Accessed 1 Dec 2007

Tress ML et al.(2007) The implications of alternative splicing in the ENCODE protein complement. Proc Natl Acad Sci USA 104:5495–5500

Tumour-Host Genomics (2007) FP6 (2007) project on tumour host genomics. http://research.med.helsinki.fi/tumorhostgenomics. Accessed 1 Dec 2007

UJF (2007) University Joseph Fourier – Grenoble. http://www.ujf-grenoble.fr. Accessed 1 Dec 2007

UK-Sysbio (2007) UK's integrative systems biology: BSSRC + EPSRC program http://149.155.200.17/about/gov/panels/isb_intro.html#top. Accessed 1 Dec 2007

Underwood JCE (2002) General and systematic pathology, 3rd edn. Churchill Livingstone, Edinburgh

UniProt (2007) Universal protein resource. http://www.ebi.ac.uk/uniprot/. Accessed 1 Dec 2007

VALAPODYN (2007) Validated predictive dynamic model of complex intracellular pathways related to the cell death and survival. http://www.valapodyn.eu. Accessed 1 Dec 2007

Valdar W, Solberg LC, Gauguier D, Burnett S, Klenerman P, Cookson WO, Taylor MS, Rawlins JN, Mott R, Flint J (2006) Genome-wide genetic association of complex traits in heterogeneous stock mice. Nat Genet 38(8):879–887

Vander A, Sherman J, Luciano D (2001) Human physiology, McGraw-Hill, New York 8th edn.

Vanvossel A (ed) (2005) Major diseases research – projects funded under the Sixth Framework Programme (2002–2005). European Commission, Brussels. ftp://ftp.cordis.europa.eu/pub/lifescihealth/docs/major_catalogue_complet.pdf. Accessed 1 Dec 2007

Variome (2007) The Human Variome Project. http://www.variome.org/. Accessed 1 Dec 2007

VIDA (2007) Virus database of homologous protein families. http://www.biochem.ucl.ac.uk/bsm/virus_database/VIDA.html. Accessed 1 Dec 2007

ViralDAS (2007) Viral DAS (2007) server. http://viraldas.bioinf.mpi-inf.mpg.de/index.php. Accessed 1 Dec 2007

ViRgil (2007a). BioSapiens-ViRgil workshop on Bioinformatics for Viral Infections. http://workshop2005.bioinf.mpi-sb.mpg.de/. Accessed 1 Dec 2007

ViRgil (2007b) Combatting viral resistance to treatments. http://www.virgil-net.org/. Accessed 1 Dec 2007

Virtual-Wormbase (2007) The virtual worm database. http://celegans.sh.se. Accessed 1 Dec 2007

Wälchli S, Espanel X, Harrenga A, Rossi M, Cesareni G, Huijsduijnen RH van (2004) Probing protein-tyrosine phosphatase substrate specificity using a phosphotyrosine-containing phage library. J Biol Chem 279(1):311–318

Weinberg RA (2007) The biology of cancer. Garland, New York

Weinberg-Contents (2007) Table of contents of biology of cancer by Weinberg. http://www.garlandscience.co.uk/textbooks/0815340788/pdf/TableofContents.pdf. Accessed 1 Dec 2007

Wellcome Trust (2007) The world's largest medical research charity funding research into human and animal health. http://www.wellcome.ac.uk/. Accessed 1 Dec 2007

Whatizit (2007) Text processing system. http://www.ebi.ac.uk/webservices/whatizit/info.jsf. Accessed 1 Dec 2007

Wolpert L, Beddington R, Jessell T, Lawrence P, Meyerowitz E, Smith J (2001) Principles of development, 2nd edn. Oxford University Press, Oxford

Woolf N (2000) Cell, tissue and disease – the basis of pathology. Saunders, Philadelphia

Wormbase (2007) The biology and genome of *C. elegans*. http://www.wormbase.org/. Accessed 1 Dec 2007

WTCCB (2007) Welcome Trust Centre for Cell Biology. http://www.wcb.ed.ac.uk/. Accessed 1 Dec 2007

WTCCC (2007) The Wellcome Trust Case Control Consortium. http://www.wtccc.org.uk/. Accessed 1 Dec 2007

WTSI (2007) The Wellcome Trust Sanger Institute. http://www.sanger.ac.uk/. Accessed 1 Dec 2007

YSBN (2007) Yeast Systems Biology Network. http://www.ysbn.eu. Accessed 1 Dec 2007

YSBN-tools (2007) Links to systems biology tools. http://www.ysbn.eu. Accessed 1 Dec 2007

Zeggini E et al.(2007) Replication of genome-wide association signals in UK samples reveals risk loci for type 2 diabetes. Science 316(5829):1336–1341.

Index

MIX
Papier aus verantwortungsvollen Quellen
Paper from responsible sources
FSC® C105338

If you have any concerns about our products,
you can contact us on
ProductSafety@springernature.com

In case Publisher is established outside the EU,
the EU authorized representative is:
Springer Nature Customer Service Center GmbH
Europaplatz 3, 69115 Heidelberg, Germany

Printed by Libri Plureos GmbH
in Hamburg, Germany